APPLIED MATHEMATICS

FOR THE PETROLEUM AND OTHER INDUSTRIES

Third Edition

Edited by
Mildred Gerding

Published by
PETROLEUM EXTENSION SERVICE
Division of Continuing Education
The University of Texas at Austin
Austin, Texas

in cooperation with
AMERICAN PETROLEUM INSTITUTE
Production Department
Dallas, Texas

1985

First Edition published 1948. Third Edition 1985
Second Impression 1988
Printed in the United States of America

Catalog No. 1.60030
ISBN 0-88698-085-2

CONTENTS

FOREWORD v

ACKNOWLEDGMENTS vii

1. THE NUMBER SYSTEM 1
   - Objectives 1
   - Whole Numbers 2
   - Common Fractions 12
   - Mixed Numbers 18
   - Decimal Fractions 22
   - Roots and Powers 30
   - Self-Test 34

2. THE CALCULATOR 39
   - Objectives 39
   - Choosing a Calculator 40
   - Understanding Calculator Functions 44
   - Operating the Calculator 47
   - Self-Test 55

3. NUMBER RELATIONS 57
   - Objectives 57
   - Percentage 58
   - Ratio and Proportion 62
   - Mean (Average), Median, and Mode 66
   - Tables and Graphs 69
   - Self-Test 78

4. SOME PHYSICAL QUANTITIES AND THEIR MEASUREMENT 83
   - Objectives 83
   - Length, Area, and Volume 86
   - Time and Temperature 96
   - Mass and Related Derived Quantities 101
   - Electricity 108
   - Self-Test 115

5. PRINCIPLES OF ALGEBRA 119
   - Objectives 119
   - Algebraic Expressions 120
   - Equations 126
   - Formulas 129
   - Self-Test 138

6. PRACTICAL GEOMETRY 143
Objectives 143
Plane Figures 144
Solid Figures 155
Geometric Constructions 168
Self-Test 181

7. TRIGONOMETRY 187
Objectives 187
Right Triangle Trigonometry 188
Oblique Triangle Calculations 196
Self-Test 202

8. ADVANCED OIL INDUSTRY APPLICATIONS 207
Objectives 207
Drilling 208
Production 222
Pipeline 229
Refining 236
Self-Test 241

APPENDICES
A. Commonly Used Formulas 243
B. Reference Tables 247
C. Conversion Factors 252

ANSWER KEY 259

FOREWORD

When the first edition of this manual was published in 1948, the objective was to supply a basic reference work to be used principally for review by field employees of the oil and gas industry. The petroleum industry had a definite need for a comprehensive work that emphasized the application of math to field and shop work.

During the intervening years, distribution has broadened to include students and employees in other industrial areas. This use and the widespread availability of the electronic calculator necessitated a change in the structure of the book. This new edition is geared toward its use as a text as well as a reference.

It is strongly recommended that students acquire and use a handheld calculator except during the first chapter. It is important that this chapter be completed without the use of a calculator, so that students and workers are able to perform basic arithmetic operations when no calculator is at hand.

When used primarily as a refresher course in elementary mathematics, it is assumed that the user will refer to those chapters where review is needed. Practice problems and self-tests are included to help the self-study student test his mathematics skills in the desired areas.

As always, Petroleum Extension Service (PETEX) will be most grateful to hear from those who find errors or erroneous content, so that these can be corrected in the next printing.

Curtis Kruse, Director
Petroleum Extension Service

ACKNOWLEDGMENTS

Sincere appreciation is extended to the many people who have contributed their time and effort toward keeping *Applied Mathematics* abreast of the needs of the industry. Reflected in the new edition are many suggestions offered by Dois D. Dallas, Petroleum Technology Department, Tanana Valley Community College, Fairbanks, Alaska.

The American Petroleum Institute, Production Department (then the Division of Production), sponsored the first oil field mathematics books in the early 1930s and subsequently cooperated with PETEX by providing reviewers for the first and second editions of this PETEX manual. Individuals who contributed to this third edition in cooperation with the API Production Department include the following:

Tommy Bicknell, Sedco, Inc.
R. G. Carson, Exxon Company, U.S.A.
Charlie Smith, Conoco, Inc.
Howard Swaim, Conoco, Inc.
Lejeune Wilson, Sedco, Inc.

Special thanks go to Bruce R. Whalen, PETEX training specialist emeritus, who wrote the measurement chapter and served as consultant throughout the book. Other PETEX staff members who contributed to this revision are Ron Baker, Gale Hathcock, Celina Henderson, Jodie Leecraft, Jeanette Paxson, and Marion Vose.

1 THE NUMBER SYSTEM

Objectives

Upon completion of chapter one, the student, without the aid of a calculator, will be able to—

1. Convert numbers that are written out into digits in their proper places.
2. Convert numbers from one unit into another given unit.
3. Add and subtract whole numbers.
4. Multiply and divide whole numbers.
5. Determine whether a word problem requires addition, subtraction, multiplication, or division.
6. Find the least common denominator of a group of fractions.
7. Reduce a common fraction to its lowest terms.
8. Add and subtract common fractions.
9. Multiply and divide common fractions.
10. Add and subtract mixed numbers.
11. Multiply and divide mixed numbers.
12. Add and subtract decimal fractions.
13. Multiply and divide decimal fractions.
14. Convert common fractions into decimal fractions and decimal fractions into common fractions.
15. Find the square root of a number.
16. Calculate the quantity of a number squared or designated to some other power.

Primitive man could count to ten using his ten fingers. Then, he could start over and count to ten again. When he counted to ten, ten times, he had one hundred. Later all numbers from one to ten were given symbols or figures to represent them. These figures were called *Arabic* because of their origin in the Middle East. In contrast, another system of symbols in use at that time was the Roman number system. Roman numerals are now useful mainly for chapter headings in books, the hours on clocks, and copyright dates for movies.

The Arabic figures are 0, 1, 2, 3, 4, 5, 6, 7, 8, and 9. If the Arabic system contained nothing more than these figures, it might be worthwhile just as a means of saving time in writing; that is, 3 is easier to write than the Roman III. But this set of symbols is just the beginning. The really revolutionary aspect of the Arabic system is that the value of each symbol is determined by its placement; that is, a given figure's position determines the value of that number. For example:

thousands	hundreds	tens	ones		
0	0	0	5	=	five, or 5 ones
0	0	5	0	=	fifty, or 5 tens
0	5	0	0	=	five hundred, or 5 hundreds
5	0	0	0	=	five thousand, or 5 thousands.

With this ingenious number system, it is possible to write any whole number by combining the ten symbols (0, 1, 2, 3, 4, 5, 6, 7, 8, 9). This Arabic system based on ten is the conventional system of numbers in use throughout the world today. It is generally called the decimal system from the Latin word for ten, *decem.*

Whole Numbers

Table 1.1 shows the decimal system of using Arabic figures to make whole numbers, both large and small. This number system is divided from right to left into *units* (or *ones*), *thousands, millions, billions,* and *trillions.* Each of the larger divisions is again divided from right to left into *ones, tens,* and *hundreds.* The divisions indicate position value and make it easier to read large numbers. In the conventional system, commas are used to show the divisions between the units, thousands, millions, billions, and trillions.

It is also important to note that 1 *ten* is the same as 10 *ones,* and 1 *hundred* is the same as 10 *tens.* Continuing this process, 1 *thousand* is the same as 10 *hundreds* and 1 *million* is the same as 1,000 *thousands.* This conversion process is different from recognizing that a digit is in a given position. These conversions are important in the subtraction operation.

Table 1.1 also shows how to read and write whole numbers. For example, line nine in the table is read as "five hundred fifty-five million, five

Table 1.1
Decimal System of Numbers

Number Written in Arabic Symbols	Number Placement															Number Spoken or Spelled Out
	Trillions			Billions			Millions			Thousands			Unit/Ones			
	100	10	1	100	10	1	100	10	1	100	10	1	100	10	1	
5															5	five
42														4	2	forty-two
340													3	4	0	three hundred forty
6,088												6	0	8	8	six thousand eighty-eight
44,307											4	4	3	0	7	forty-four thousand, three hundred seven
529,600										5	2	9	6	0	0	five hundred twenty-nine thousand, six hundred
1,000,000									1	0	0	0	0	0	0	one million
25,202,019								2	5	2	0	2	0	1	9	twenty-five million, two hundred two thousand, nineteen
555,555,555							5	5	5	5	5	5	5	5	5	five hundred fifty-five million, five hundred fifty-five thousand, five hundred fifty-five
2,250,000,000						2	2	5	0	0	0	0	0	0	0	two billion, two hundred fifty million
30,000,000,000					3	0	0	0	0	0	0	0	0	0	0	thirty billion
300,000,000,000				3	0	0	0	0	0	0	0	0	0	0	0	three hundred billion
1,064,038,000,000			1	0	6	4	0	3	8	0	0	0	0	0	0	one trillion, sixty-four billion, thirty-eight million
10,454,300,000,000		1	0	4	5	4	3	0	0	0	0	0	0	0	0	ten trillion, four hundred fifty-four billion, three hundred million
254,100,000,000,000	2	5	4	1	0	0	0	0	0	0	0	0	0	0	0	two hundred fifty-four trillion, one hundred billion

hundred fifty-five thousand, five hundred fifty-five," showing that each 5 is ten times as great as the 5 immediately to its right; hence, moving any number one place to the left is the same as multiplying it by 10.

NOTE: The instructions and examples given in this chapter assume the reader is ***not*** *using a calculator.*

ADDITION

Addition is the process of uniting two or more numbers or groups of objects of the same kind. The number obtained by adding is called the *sum*. The plus symbol (+) is the operational sign of addition.

It is important to remember that only like quantities can be added; that is, it would not be possible to add 5 dollars and 3 bolts to get 8 dollars or 8 bolts. Nor is it possible to add 6 pounds of grease and 90 gallons of oil to get 96 pounds or 96 gallons of lubricant. The unit of measurement must be the same.

The first step in addition is to arrange the numbers uniformly in vertical columns; that is, the units digit of each number in the right-hand or units column; the *tens* digit of each number in the next column to the left, or *tens* column; and so forth. Regardless of the sizes and lengths of the numbers, the last digit to the right of each number will be in a vertical column.

Add the numbers in the right-hand column. If the sum can be expressed by one digit, write it under the column. If the sum contains more than one digit (10 or more), write the right-hand digit of the sum under the column and add the remaining digit or digits to the top of the next column to the left.

Example Problem: Add the following numbers: 8,145; 234; and 56.

Solution:

```
     1 1
   8,1 4 5
     2 3 4
 +     5 6
 ---------
   8,4 3 5
```

Add the right-hand column first. Write or "put down" 5 and carry the 1 to add to the next column. Add the next column; put down 3 and carry 1. Add the third column; put down 4. Add the last column; put down 8. The sum is 8,435.

Checking addition can be done by either of two methods: (1) by adding the column numbers in reverse sequence or (2) by adding column totals separately. The first method merely changes the mental addition sequence. If a difference occurs in the totals, both sequences should be added again to locate the error. In the second method, often used in adding long columns of figures, each column total is moved one place to the left of the preceding total; then these totals are added.

Example Problem: Check the addition of the preceding problem by adding column totals.

Solution:

```
   8,1 4 5
     2 3 4
 +     5 6
 ---------
       1 5
     1 2
     3
 + 8
 ---------
   8,4 3 5
```

SUBTRACTION

Subtraction is the process of taking an amount away from another amount or taking one number away from another number and arriving at a third number. The number that is subtracted is called the *subtrahend,* the number from which it is taken is known as the *minuend,* and the number that represents the results or answer of the process of subtraction is called the *remainder,* or *difference.* The minus symbol (−) is the operational sign for subtraction. As with addition, only like quantities can be subtracted. Also, numbers should be arranged uniformly in vertical columns with the units digit of each number in the right-hand column, the tens digits in the next column, and so on.

Example Problem: Subtract 83 from 597.

Solution:

$$\begin{array}{rl} 5\,9\,7 & \text{(minuend)} \\ -\quad 8\,3 & \text{(subtrahend)} \\ \hline 5\,1\,4 & \text{(difference, or remainder)} \end{array}$$

Beginning with the right-hand column, decide what number must be added to the 3 in the subtrahend to result in 7 in the minuend. Write this number, 4, under the line, continue with each column until the remainder is complete.

Example Problem: 273 minus 186, or 273 − 186 = ?

Solution:

$$\begin{array}{r} 1\;16\;\; \\ \not{2}\,\not{7}\,{}^{1}3 \\ -\;1\,8\,6 \\ \hline 8\,7 \end{array}$$

The first step is to subtract 6 from 3; however, 6 is bigger than 3. So, borrow a 10 from the tens place, reducing the 7 tens to only 6 tens and bringing the borrowed 10 over to the ones place. When you bring the 10 to the ones place, you add it to what is already there, making the 3 a 13. Now subtract the 6 from 13 to get 7. Again, in order to subtract the 8 from the remaining 6, you have to enlarge the 6 by borrowing a 100 from the hundreds place and adding it to the 6 to make it 160 and leaving 1 instead of 2 in the hundreds column. Now 8 from 16 leaves 8, and 1 from 1 leaves 0, resulting in the answer 87.

It is helpful when subtracting to keep track of borrowing by running slashes through changed numbers and writing the new number above the changed number as shown.

Subtraction can be checked by adding the subtrahend and the difference:

$$\begin{array}{r} {\scriptstyle 1\,1} \\ 186 \\ +87 \\ \hline 273 \end{array}$$

If the sum equals the minuend, the subtraction was done correctly.

MULTIPLICATION

Multiplication is the process of adding one number to itself a specified number of times. The number which is multiplied, or increased in multiples, is called the *multiplicand.* The *multiplier* is the number of times the multiplicand is to be added to itself. Thus, 451 times 25 means that 451 (the multiplicand) is added to itself 25 (the multiplier) times. The result obtained by multiplying one number by another is called the *product.* The sign of multiplication is × or • and is called *times.* For example, 3 × 4 or 3 • 4 is read as "three times four."

Multiplication shortens the process of addition. For instance, 3 + 3 + 3 + 3 is the same as 3 × 4, and both equal 12. A multiplication table (table 1.2) has been devised to aid in this shortening process. To find the answer of 3 × 4 in the table, locate the multiplicand, 3, in the horizontal row across the top of the table and the multiplier, 4, in the left-hand vertical column. Follow these two numbers horizontally and vertically until they meet at 12, the product. In order to successfully multiply without a table or calculator, it is essential that the multiplication table be thoroughly and accurately learned.

Table 1.2
Multiplication Table

1	**2**	**3**	**4**	**5**	**6**	**7**	**8**	**9**	**10**	**11**	**12**
2	4	6	8	10	12	14	16	18	20	22	24
3	6	9	12	15	18	21	24	27	30	33	36
4	8	12	16	20	24	28	32	36	40	44	48
5	10	15	20	25	30	35	40	45	50	55	60
6	12	18	24	30	36	42	48	54	60	66	72
7	14	21	28	35	42	49	56	63	70	77	84
8	16	24	32	40	48	56	64	72	80	88	96
9	18	27	36	45	54	63	72	81	90	99	108
10	20	30	40	50	60	70	80	90	100	110	120
11	22	33	44	55	66	77	88	99	110	121	132
12	24	36	48	60	72	84	96	108	120	132	144

It should be borne in mind that the multiplier is always an abstract number while the multiplicand may be either abstract or concrete and that the product is the same quantitative unit as the multiplicand. It is impossible to multiply 8 horses by 4 goats because both are concrete numbers. In like manner, it is impossible to multiply dollars by men, but it *is* possible to multiply dollar rate of pay per man (concrete) by the number of men working (abstract) and get as the product total dollars paid (concrete). Thus, $46 × 3 men = $138.

Example Problem: Multiply 232 by 3, or 232 × 3 = ?

Solution:

```
   2 3 2  (multiplicand)
×      3  (multiplier)
---------
   6 9 6  (product)
```

Starting with the right-hand number 2 in the multiplicand, multiply by 3, the multiplier, to get the product 6. Put the 6 under the line. Proceed in the same way with each succeeding figure of the multiplicand, working from right to left. The resulting product is 696.

As the multiplier becomes larger, the multiplication process gets more involved and placement becomes very important. When the multiplier contains a zero, proper placement of the partial products becomes even more vital. It is also important to remember that any number multiplied by 0 is 0, and any number multiplied by 1 is the same as the number.

Example Problem: Multiply 25 × 16.

Solution:

```
    3
    2 5  (multiplicand)
  × 1 6  (multiplier)
---------
  1 5 0  (partial product)
+ 2 5    (partial product)
---------
  4 0 0  (product)
```

As the first step, multiply 6 × 5, giving 30, a number with two figures. Enter the 0 below the line and carry the 3 to the top of the next column. Then multiply 6 × 2, giving 12; add the carried 3 to the 12, resulting in 15, which is entered to the left of the zero. The next step is to multiply the number 25 by the 1 in the multiplier to get 25. This number is entered under the partial product 150 *one place to the left.* Add these two partial products to obtain the product 400.

Example Problem: You have 162 joints of 30-foot drill pipe. How many feet can be drilled with this supply?

Solution: In order to find the number of feet that can be drilled, you will have to multily 162 times 30.

```
    1 6 2   joints   (multiplier)
×     3 0   feet     (multiplicand)
-------------
× 4,8 6 0   feet     (product)
```

Since the last figure in the multiplicand is 0, it is placed with the 0 one place to the right of the multiplier and is put under the multiplier instead of over it for convenience' sake. The 0 is brought down and the 3 is then multiplied times the 162 in the usual manner.

Since 0 times any number remains 0, it is not necessary to multiply out and put a row of zeros in the answer. The important thing is that the 0 be brought down as a place holder. When the multiplier contains one or more zeros between two other numbers, the same principle will apply. The partial products should be moved to the left two places instead of one for each 0.

Example Problem: An oil company conducted a 105-hour school for 345 of its employees. How many man-hours of training did the company provide in this school?

Solution: Each man received 105 hours of training, thus 345 × 105 will give the number of man-hours.

```
   3 4 5
 × 1 0 5
 -------
   1 7 2 5
 3 4 5 0
 -------
 3 6,2 2 5   man-hours of training provided
```

Checking multiplication can be done by exchanging the multiplier and multiplicand and multiplying again. It can also be checked by dividing the multiplicand into the product to obtain the multiplier, or to divide the multiplier into the product to obtain the multiplicand. (See following section on division.)

DIVISION

Division is the reverse of multiplication. It is finding out how many times one number is contained in another number. To divide 8 by 4 means to find how many times 4 is contained in 8 or, stated another way, what number is multiplied by 4 to result in 8. The answer to this problem is, of course, 2.

One symbol used to indicate division is ÷ and is read "divided by." When the division sign (÷) appears between two numbers, it indicates that the first number is to be divided by the second. For example, 16 ÷ 4 is read "sixteen divided by four" and means what number will equal 16 if multiplied by 4.

The number to be divided is called the *dividend,* and the number which does the dividing is called the *divisor.* The result obtained by division is called the *quotient.*

There are two methods of division, short division and long division. Short division is used when the divisor contains only one digit.

Example Problem: Divide 345 by 3, or what number multiplied by 3 equals 345?

Solution: Write down the numbers in the following manner and use the short division method.

```
             1 1 5  (quotient)
(divisor)  3/3 4 5  (dividend)
```

Divide the first or left-most figure in the dividend by 3, obtaining 1 (3 is contained in 3 one time). Write this 1 above the line over the 3; then divide the next number 4 by 3. Note that 3 is contained in 4 one time, but 1 is left over. So place 1 in the quotient above the 4 and mentally place the 1 remaining in front of the 5, making it 15. Then divide 15 by 3, obtaining 5 as the final figure in the quotient.

Since division is the reverse of multiplication, it is checked by multiplying the answer or quotient by the divisor. The product must be the same as the dividend. For example:

$$\begin{array}{r} 115 \\ \times \quad 3 \\ \hline 345 \end{array}$$

Example Problem: 552 ÷ 23 = ?

Solution: Using long division, write down the numbers in the following manner.

```
     2 4
2 3/5 5 2
    4 6
   -----
    9 2
    9 2
   -----
      0
```

Since 23 is not contained in 5, determine how many times it is contained in 55. Note that it is contained two times, so write 2 above the second 5. Then multiply this 2 times 23 to get 46. Write 46 under the 55 and subtract, getting 9. Bring down the remaining digit 2 from the dividend and place beside the 9, making 92. Now divide 92 by 23. By trial it is found that 23 is contained exactly four times in 92, so 4 is placed to the right in the quotient. Multiply 4 times 23 to get 92; subtract 92 from 92 to get 0, or a remainder of nothing. This shows that the process is complete and that 552 divided by 23 is equal to 24, or that 23 is contained in 552 exactly 24 times.

Many problems involving division of whole numbers do not work out evenly but have a number left over, called a *remainder,* when the divisor is not contained in the dividend an exact number of times. For instance, 8 is not contained in 45 an exact number of times:

$$45 \div 8 = 5 \text{ with a remainder of } 5$$

In long division, the remainder is written to the right of the answer preceded by an *r.*

Example Problem: 4,868 ÷ 30 = ?

Solution:

```
      1 6 2 r 8
3 0 / 4,8 6 8
    -3 0
     1 8 6
    -1 8 0
         6 8
         6 0
           8
```

PRACTICE PROBLEMS

1. Convert the following numbers into the units given.

   a. 5 millions is the same as __________ thousands. __________

   b. 11 thousands is the same as __________ hundreds. __________

   c. 3 tens is the same as __________ ones. __________

   d. 1 hundred is the same as __________ tens. __________

2. Write the Arabic figures for the numbers written out.

   a. There are *forty-two* gallons in a barrel of oil. __________

b. Total footage drilled in 1981 was *three hundred sixty-one million.* ____________

c. At the beginning of 1980 the U.S. proved crude oil reserves totaled *twenty-seven billion, fifty-one million, two hundred and eighty-nine thousand* barrels. ____________

d. The 1981 marketed production of natural gas in the United States was *twenty trillion, four hundred billion, five hundred and two million* cubic feet. ____________

3. Write out the numbers for the Arabic figures in the blanks below.

a. The average depth of wells drilled in 1981 was *4,600* feet. ____________

b. The number of wells drilled in the U.S. in 1981 was *78,000.* ____________

c. Crude oil production in the U.S. in 1981 averaged *8,600,000* barrels per day. ____________

d. Fossil fuel imports by the U.S. in 1981 totaled *$80,740,000,000.* ____________

4. Work the following problems.

a. 352 + 569 + 32 = ____________

b. 2,832 ÷ 12 = ____________

c. 1,950 × 657 = ____________

d. 7,408,732 − 923,351 = ____________

5. Work the following word problems, using addition, subtraction, multiplication, or division as needed.

a. A welding company has 36 welders working at the same rate per month. The total monthly payroll for these employees is $88,200. How much does each welder make per month? ____________

How much per year? ____________

b. A truck averages 55 miles per hour. How many hours of driving time will be needed to make a 935-mile trip? ____________

c. A drilling crew has 35 joints of 30-foot drill pipe and a 30-foot kelly down in the hole. What is the approximate depth of the well? ____________

d. If a well is 6,500 feet deep, and 563 yards of the well has casing in it, how many more feet of casing are needed to hit bottom? (Remember, 3 feet = 1 yard.) ____________

Common Fractions

Although cumbersome and time-consuming to use in computations, common fractions are inescapable under the industrial practices followed in the United States, England, and some other countries. In American industry, fractions are used chiefly to size manufactured products and the machine tools used in manufacturing. Stocks and bonds are quoted in the security markets in dollars and fractions of a dollar.

In industry, dimensions of bolts, nuts, tubular goods, machine parts, tanks, and similar equipment require the use of common fractions. Some specific uses of fractions include the following:

1. Pipe diameters are sized in halves, fourths, and eighths of an inch.
2. Flow beans are sized in thirty-seconds and occasionally sixty-fourths of an inch.
3. Wire rope is sized in halves, fourths, eighths, and sixteenths of an inch.
4. Depth of oil in a tank is measured in halves and fourths of an inch.
5. Shaft diameters, vessel openings and wall thickness, cylinder bores, and compressor and engine strokes, are usually measured in halves, fourths, or eighths of an inch.
6. Royalty—that portion of the crude production due the landowner—is normally one-eighth of the production but may be any other fraction agreed upon.

Oil itself is not described in common fractions. It is gauged in feet, inches, and fractions of an inch, and these measurements are immediately converted to barrels and decimal fractions of a barrel (or to metric tons and decimal fractions of a metric ton for international commerce).

The most precise measurement using a common fraction is in the machine shop where the machinist may use sixty-fourths of an inch. For more precise work, he uses a micrometer caliper based on the decimal fraction system. When common fractions are used in the oil industry, they are almost invariably fractions of an inch.

To add, subtract, multiply, and divide common fractions, it is necessary to understand what a fraction is and how it is dealt with in Arabic figures. Fractions describe parts of whole things. In the fraction 3/4, the 3 is known as the *numerator,* and the 4 is the *denominator.* The denominator indicates how many parts something whole is divided into, or how many parts it takes to make a whole. The numerator indicates how many of the total number of parts are being considered. For example, 1 inch is 1/12 of a foot, or 1 of the 12 inches in a foot; 30 minutes is 1/2 of an hour because 30 minutes is one of the two parts of an hour.

The fraction 5/5 means that something is divided into five parts (a pie, for example) and that all five parts put together make one whole. Likewise, two halves equal one whole, four quarters (or fourths) equal one whole, and so forth. Any fraction that has the same number in its numerator and denominator is the same as the whole number 1.

When working with fractions, a key principle is that the numerator and denominator can be multiplied or divided by the same number without changing the value of the fraction. For example, $1/2$ is the same as $2/4$, $4/8$, $8/16$, and so forth.

ADDITION

Before fractions can be added, they must have a *common denominator;* that is, all fractions to be added must have the same denominator. To add two or more fractions with the same denominator, simply add the numerators and place the sum over the common denominator. The denominator does not change. For example:

$$\frac{1}{8} + \frac{5}{8} = \frac{6}{8}.$$

When fractions have different denominators, it is necessary to find the *least common denominator* (LCD), that is, the smallest number that all of the denominators will divide into evenly. Often the largest denominator of the fractions will also be the LCD. For example, to add $1/4$ and $3/16$, the 16 would be used as the LCD. In order to express $1/4$ in sixteenths, both the numerator 1 and the denominator 4 must be multiplied by a number that will make the new denominator 16. Dividing 4 into 16 gives the number 4, so 4 is multiplied times 1 and 4 to give 4 and 16, respectively, making the fraction $4/16$. Hence,

$$\frac{1}{4} + \frac{3}{16} = \frac{4}{16} + \frac{3}{16} = \frac{7}{16}.$$

Remember, multiplying the numerator and denominator by the *same* number does not change the value of the fraction.

Example Problem: Add $3/4$, $7/8$, and $13/16$.

Solution: Set up the problem and find the LCD.

$$\frac{3}{4} + \frac{7}{8} + \frac{13}{16} = ?$$

The LCD is 16, since 4 and 8 will both divide evenly into 16. Converting $3/4$ into sixteenths,

$$\frac{3}{4} = \frac{3 \times (16 \div 4)}{4 \times (16 \div 4)} = \frac{3 \times 4}{4 \times 4} = \frac{12}{16}.$$

Similarly,

$$\frac{7}{8} = \frac{7 \times (16 \div 2)}{8 \times (16 \div 2)} = \frac{7 \times 2}{8 \times 2} = \frac{14}{16}.$$

Now the problem can be set up with like denominators and can be solved by adding the numerators as

$$\frac{12}{16} + \frac{14}{16} + \frac{13}{16} = \frac{39}{16}.$$

Since 39 is larger than 16, it is evident that the final answer is more than 1. To reduce this fraction to a whole number and a fraction, divide 39 by 16:

```
   2 r 7
16/39
   32
   --
    7
```

The remainder 7 becomes the numerator, and 16 remains the denominator, making the fraction 7/16. Thus, the answer is 2 7/16.

SUBTRACTION

Subtracting common fractions is based on the same principles as adding common fractions. The first step in subtraction is to find a common denominator. Then one numerator is subtracted from the other, and the difference is written over the common denominator.

Example Problem: Subtract 5/8 from 3/4.

Solution: Using 8 as the LCD, convert 3/4 to eighths.

$$\frac{3}{4} = \frac{3 \times (8 \div 4)}{4 \times (8 \div 4)} = \frac{3 \times 2}{4 \times 2} = \frac{6}{8}.$$

Then subtract numerators,

$$\frac{6}{8} - \frac{5}{8} = \frac{1}{8}.$$

Example Problem: Find the difference between 15/16 of an inch and 19/64 of an inch.

Solution: Find the least common denominator.

$$\frac{15}{16} = \frac{15 \times (64 \div 16)}{16 \times (64 \div 16)} = \frac{15 \times 4}{16 \times 4} = \frac{60}{64}.$$

Then subtract numerators.

$$\frac{60}{64} - \frac{19}{64} = \frac{41}{64}.$$

Sometimes the resulting number can be reduced to a smaller fraction. This is done only if dividing both the numerator and denominator by the same number results in even numbers. For example, 60/64 can be reduced back to 15/16 by dividing by 4. However, 41/64 cannot be reduced, since no number other than 1 will divide evenly into 41. This operation, called *reducing to the lowest terms,* is especially necessary in the multiplication of fractions.

MULTIPLICATION

When one fraction is multiplied by another, the numerators are multiplied together to get a new numerator, and the denominators are multiplied together to provide a new denominator. The new fraction that is the product should be reduced to its lowest terms. It should be divided by numbers contained in both the numerator and denominator until these terms are as small as it is possible to reduce them. In the fraction 15/25, for example, dividing both terms by 5 results in 3/5. Since no number other than 1 will divide evenly into 3 and 5, the fraction 3/5 is the lowest term.

Example Problem: Multiply 3/4 by 5/7.

Solution:

$$\frac{3}{4} \times \frac{5}{7} = \frac{3 \times 5}{4 \times 7} = \frac{15}{28}.$$

Example Problem: 3/16 × 4/5 × 1/2 = ?

Solution:

$$\frac{3 \times 4 \times 1}{16 \times 5 \times 2} = \frac{12}{160} = \frac{3}{40}.$$

12/160 is reduced to its lowest terms by dividing it by 4.

Cancellation is a method of shortening the multiplication of fractions. It entails dividing any numerator and any denominator by the same number, then multiplying the new, smaller numbers.

Example Problem: 5/12 × 3/15 × 3/8 = ?

Solution: Cancellation here saves time because there are numbers in the numerators and denominators that can be divided by the same number.

$$\frac{\overset{1}{\cancel{5}}}{\underset{4}{\cancel{12}}} \times \frac{\overset{1}{\cancel{3}}}{\underset{\underset{1}{\cancel{3}}}{\cancel{15}}} \times \frac{\overset{1}{\cancel{3}}}{8} = \frac{1}{4} \times \frac{1}{1} \times \frac{1}{8} = \frac{1}{32}.$$

First, the numerator 5 (first fraction) and the denominator 15 (second fraction) are both divided by 5, leaving 1 and 3, respectively. Then the numerator 3 (second fraction) and the denominator 12 (first fraction) are divided by 3, leaving 1 and 4. The numerator 3 (third fraction) and the new denominator 3 (second fraction) cancel each other, leaving 1 and 1. Now multiply $1 \times 1 \times 1$ to get 1 in the numerator, and $4 \times 1 \times 8$ to get 32 in the denominator.

DIVISION

Dividing 12 by 3 is the same as multiplying 12 by $1/3$. When division is indicated, the process can be carried out by interchanging the numerator and denominator of the divisor and multiplying. In other words, the fraction doing the dividing is inverted and used as a multiplier.

Example Problem: Divide $3/16$ by $9/64$.

Solution:

$$\frac{3}{16} \div \frac{9}{64} = \frac{\overset{1}{\cancel{3}}}{\underset{1}{\cancel{16}}} \times \frac{\overset{4}{\cancel{64}}}{\underset{3}{\cancel{9}}} = \frac{4}{3} = 1\tfrac{1}{3}.$$

PRACTICE PROBLEMS

1. Convert the following fractions to the denominators given.

   a. $3/32$ to 64ths __________

   b. $3/4$ to 16ths __________

   c. $56/64$ to 16ths __________

   d. $14/16$ to 8ths __________

2. Reduce the following fractions to the lowest common denominators.

   a. $1/2$, $1/8$, $3/4$ ____________

   b. $1/8$, $1/16$, $7/32$ ____________

   c. $7/8$, $19/64$, $11/32$, $17/64$, $1/16$, $62/64$ ____________

   d. $31/32$, $15/16$, $3/64$ ____________

3. Reduce the following fractions to their lowest terms.

   a. $32/64$ ____________

   b. $56/64$ ____________

   c. $4/32$ ____________

   d. $10/12$ ____________

4. Add or subtract the following fractions.

   a. $1/2 + 3/4 + 1/4 =$ ____________

   b. $59/64 + 3/32 + 9/16 + 7/64 + 9/64 + 1/8 + 7/16 + 59/64 =$ ____________

   c. $5/16 - 3/32 =$ ____________

   d. $26/32 - 9/64 =$ ____________

5. Multiply or divide the following fractions.

   a. $5/8 \times 9 =$ ____________

   b. $7/16 \times 5/10 =$ ____________

   c. $11/12 \div 5/6 =$ ____________

   d. $(5/6 \div 1/4) \times 2/3 \times 1/8 =$ ____________

6. A well produces 640 barrels of oil and salt water per day. If three-eighths of the production is oil, and the lease owner receives one-eighth royalty on this, how many barrels of oil will he receive royalty on daily? ____________

7. Jones has a three-eighths interest in a lease, Smith owns three-sixteenths, and White owns seven-sixteenths. If they sell the lease for $75,000, how much will each receive?
   Jones: ______
   Smith: ______
   White: ______

8. What is the difference in the diameter of two wirelines that are $7/8''$ and $9/16''$ in diameter? ______

9. A storage tank is 7′6″ high overall. The top plates are $3/16''$ thick, and the bottom plates are $1/4''$ thick. What is the inside height of the tank? ______

10. A worn pump plunger that is $15/16''$ in diameter is placed in a lathe, and three cuts are taken off it. On the first cut, $3/16''$ is removed, on the second cut $3/64''$, and on the third cut $3/8''$. What is the diameter of the plunger after these cuts are taken? ______

Mixed Numbers

A mixed number is a number composed of a whole number and a fraction. For example, $18\frac{1}{4}$, $272\frac{1}{2}$, and $5\frac{3}{4}$ are mixed numbers. A mixed number is read as if the word *and* were between the whole number and the fraction. That is, $18\frac{3}{4}$ is read as "eighteen and three-fourths."

ADDITION

To add mixed numbers, the fractions have to be changed so that they will have a common denominator. Then the whole numbers are added and the fractions are added.

Example Problem: Find the sum of $9\frac{1}{4}$ and $12\frac{1}{8}$.

Solution:

$$\begin{array}{rcr} 9\frac{1}{4} & = & 9\frac{2}{8} \\ +\ 12\frac{1}{8} & = & 12\frac{1}{8} \\ \hline & & 21\frac{3}{8} \end{array}$$

Should the fractions add up to more than 1, the whole number is carried over to the ones column and added in as a whole number.

Example Problem: Add $13\frac{3}{8}$, $17\frac{1}{8}$, $8\frac{1}{2}$, and $340\frac{1}{4}$.

Solution: First, set up the fractions with a common denominator and add the fractions. Then convert the sum of the fractions into its lowest terms.

$$\begin{aligned} 3/8 &= 3/8 \\ 1/8 &= 1/8 \\ 1/2 &= 4/8 \\ \underline{1/4} &= \underline{2/8} \\ & \quad 10/8 = 1\,2/8 = 1\,1/4 \end{aligned}$$

Then add the whole numbers, including the $1\,1/4$ in the total.

$$\begin{array}{r} 1\,1/4 \\ 13 \\ 17 \\ 8 \\ +\ 340 \\ \hline 379\,1/4 \end{array}$$

SUBTRACTION

In subtracting one mixed number from another, the fractions and the whole numbers should be considered together. This is because the fraction in the subtrahend may be larger than the fraction of the minuend, in which case it will be necessary to convert one unit in the whole number of the minuend into a fraction.

Example Problem: Subtract $3\,5/8$ from $12\,3/4$.

Solution: Change $3/4$ and $5/8$ so that both fractions have the same denominator; then subtract fractions and whole numbers.

$$\begin{array}{rcr} 12\,3/4 & = & 12\,6/8 \\ -\ 3\,5/8 & = & -\ 3\,5/8 \\ \hline & & 9\,1/8 \end{array}$$

Example Problem: Subtract $7\,5/16$ from $18\,1/8$.

Solution: Find the LCD of the fractions.

$$\begin{aligned} 1/8 &= 2/16 \\ 5/16 &= 5/16 \end{aligned}$$

It is obvious that $5/16$ cannot be taken from $2/16$, so it will be necessary to convert one unit ($16/16$) of the whole number into a fraction and add it to $2/16$. Then

$$\begin{array}{rcrl} 18\,{}^{2}\!/_{16} & = & 17\,{}^{18}\!/_{16} & \text{(minuend)} \\ -\ 7\,{}^{5}\!/_{16} & = & -\ 7\,{}^{5}\!/_{16} & \text{(subtrahend)} \\ \hline & & 10\,{}^{13}\!/_{16} & \text{(difference)} \end{array}$$

MULTIPLICATION

It is often necessary to multiply mixed numbers by whole numbers, fractions, or other mixed numbers. In order to multiply a mixed number, it has to be changed to an *improper fraction.* An improper fraction is a fraction with the numerator larger than the denominator.

To change a mixed number to an improper fraction, the whole number is multiplied by the denominator of the fraction and the product is added to its numerator. This sum is then written over the denominator to form the improper fraction. For example, to change the mixed number $3\,{}^{4}\!/_{5}$ to an improper fraction, multiply the whole number 3 times the denominator 5 to get 15. Then add the product 15 to the numerator 4, changing the numerator to 19. Thus, $3\,{}^{4}\!/_{5}$ changes to the improper fraction ${}^{19}\!/_{5}$ and can be multiplied as a common fraction. To change an improper fraction back to a whole number or a mixed number, divide the numerator by the denominator. For example, to change ${}^{15}\!/_{3}$ to a whole number, divide 15 by 3 to get 5. To change ${}^{19}\!/_{5}$ back to a mixed number:

$$5\overline{)19}^{\;3\,r\,4} = 3\,{}^{4}\!/_{5}$$

Of course, whole numbers always have a denominator of 1 when changed to a fraction.

Example Problem: Multiply $18\,{}^{1}\!/_{4}$ times $6\,{}^{1}\!/_{8}$.

Solution:

$18\,{}^{1}\!/_{4} \times 6\,{}^{1}\!/_{8} = ?$

$18\,{}^{1}\!/_{4} = {}^{73}\!/_{4}$, and $6\,{}^{1}\!/_{8} = {}^{49}\!/_{8}$.

Then $\frac{73}{4} \times \frac{49}{8} = \frac{3577}{32} = 111\,{}^{25}\!/_{32}$.

Example Problem: $18 \times 6\,{}^{1}\!/_{3} = ?$

Solution:

$$6\,{}^{1}\!/_{3} = \frac{19}{3}$$

$$\frac{\overset{6}{\cancel{18}}}{1} \times \frac{19}{\underset{1}{\cancel{3}}} = 114.$$

or

$$18 \times 6 = 108.$$

$$\frac{18}{1} \times \frac{1}{3} = \frac{18}{3} = 6.$$

$$108 + 6 = 114.$$

DIVISION

Division of mixed numbers is accomplished by changing the mixed numbers to improper fractions and dividing in the same manner as with common fractions.

Example Problem: Divide $203\frac{7}{16}$ by $24\frac{1}{8}$.

Solution:

$$203\tfrac{7}{16} = \frac{3255}{16}$$

$$24\tfrac{1}{8} = \frac{193}{8}$$

$$\frac{3255}{16} \div \frac{193}{8} = \frac{3255}{16} \times \frac{8}{193} = \frac{26040}{3088} = 8\tfrac{167}{386}$$

PRACTICE PROBLEMS

1. A workman's time on a job included $8\frac{1}{4}$, $5\frac{1}{2}$, $9\frac{3}{4}$, $7\frac{1}{4}$, $6\frac{3}{4}$, and $2\frac{1}{2}$ hours. What was the total time spent? ____________

2. Find the dimension of *A* in the sketch below. ____________

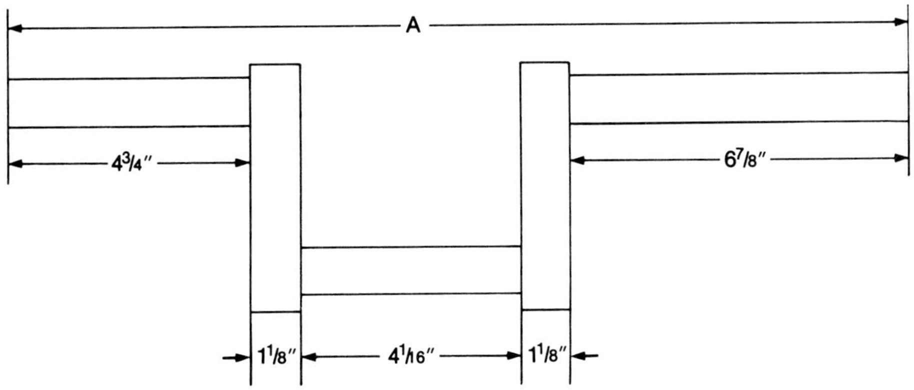

3. A roustabout cut off the following lengths of 3/4" sucker rod for pins: 5 3/16", 4 3/32", 9 1/8", and 2 1/2". What is the total length of sucker rod used? ______

4. A plant is being operated on a 40-hour week. The plant foreman's weekly time sheets for June showed that he had worked 43 1/2 hours, 42 1/4 hours, 45 3/4 hours, and 40 3/4 hours. How much overtime did he work during the month? ______

5. If the nominal inside diameter of a pipe is 5/8" and the outside diameter is 15/16", how thick is the pipe wall? (Sketching a cross section will help you visualize this problem.) ______

6. If an acetylene cutting machine is cutting at the rate of 1 3/8" every 5 seconds, how far will it cut in 2 3/4 minutes? ______

7. On a production test, a well produced 103 1/2 barrels of oil in 3 hours, 15 minutes. At this rate, what would the well produce in 24 hours? ______

8. A pipefitter is instructed to make a mark every 6 7/8", measuring from the end of a 6' pipe header. How many marks will he make on the header? ______

9. A certain number of holes are spaced equally in a straight line. If the distance between the centers of the two end holes is 66 1/4" and if the spacing from center to center of two adjacent holes is 3 5/16", how many holes are to be punched? (Note: be sure to count the beginning hole.) ______

10. If 3 pints of chemical compound are required to treat 50 barrels of cut oil, how many barrels can be treated with 8 1/4 gallons of chemical? (Note: 8 pints = 1 gallon.) ______

Decimal Fractions

Decimal numbers and fractions are being used more frequently, especially since the calculator uses decimal fractions. Casing, tubing, and drill pipe were once measured only in feet, inches, and fractions of inches. Now many operators measure with tapes divided into feet, tenths of a foot, and hundredths of a foot. Drilling fluid is weighed on the rig on a mud balance that is calibrated in pounds per gallon and tenths of a pound. In fact, decimal fractions in any computation are easier to use and more precise than common fractions.

Converting common fractions to decimal fractions and vice versa is not difficult. A decimal fraction is a fraction whose denominator is 10 or some power of 10, such as 100 or 1,000. It may be written just as a common fraction, with the numerator over the denominator, as 6/10 or 93/100 (read as six tenths or ninety-three hundredths), or it may be written as 0.6 or 0.93. To arrive at these decimal fractions, omit the denominator and indicate the denominator by the number of places the numerator occupies to the right of the *decimal point,* which is a period (.) placed to the left of the first digit in the decimal fraction. In the case of a mixed number, the decimal point is placed between the whole number and the decimal fraction.

In reading decimal fractions, one figure to the right of the decimal point indicates the number of tenths, two figures to the right of the decimal point indicate hundredths; three figures indicate thousandths, four figures indicate ten thousandths, and so on. A mixed number may be read in two ways: with the word *and* between the whole number and fraction or the word *point* where the decimal appears. Thus, 6.25 is read "six and twenty-five hundredths" or as "six point two five."

To indicate decimal fractions of the proper denomination, the right number of decimal places to the right of the decimal point is necessary. There must always be as many digits to the right of the decimal point as there would be zeros in the denominator if the decimal fraction were written as a common fraction. This is very important to remember. To write 9/100 as a decimal, there must be two digits to the right of the decimal point. Since the numerator has only one digit, a zero is placed between the decimal point and the number, making the decimal fraction 0.09.

Example Problem: Write the following decimal fractions in words: 0.25, 0.375 and 18.64.

Solution:

0.25 = twenty-five hundredths
0.375 = three hundred seventy-five thousandths
18.64 = eighteen and sixty-four hundredths

Since the U.S. monetary system is based on the decimal system, problems involving dollars and cents are among the most common decimal fraction problems encountered. Instead of reading the fractional part of a dollar as so many hundredths, it is called *cents.* For example, $8.75 is read as "eight dollars and seventy-five cents," not as "eight and seventy-five hundredths dollars."

ADDITION

To add or subtract decimal fractions or mixed decimals, they must be arranged in vertical columns *with the decimal points aligned.* Then they can be added as whole numbers.

Example Problem: Add 875.3 and 6.05.

Solution: Write the numbers so that the decimal points fall in a vertical line.

$$\begin{array}{r} 875.3 \\ +\quad 6.05 \\ \hline 881.35 \end{array}$$

Note that the decimal point in the sum is directly below the decimal points in the two figures being added.

Example Problem: Add \$35.42, \$18.78, \$19.25, and 35¢.

Solution:

$$\begin{array}{r} \$35.42 \\ 18.78 \\ 19.25 \\ .35 \\ \hline \$73.80 \end{array}$$

SUBTRACTION

Subtracting decimal fractions is handled in the same manner as addition. The decimal points are aligned, and numbers are subtracted as whole numbers. Sometimes it is necessary to add zeros to the right of a fraction in the minuend to facilitate subtraction. Placing zeros to the right of a decimal fraction does not in any way affect its value, just as placing zeros to the left of a whole number in no way affects the value of the whole number. For example, 0.27 and 0.270 have the same value, but 0.27 is its lowest term.

Example Problem: Subtract 125.873 from 340.27.

Solution:

$$\begin{array}{r} 340.270 \\ -\ 125.873 \\ \hline 214.397 \end{array}$$

MULTIPLICATION

The process of multiplying decimal fractions or mixed decimals is carried out exactly as in multiplying whole numbers, and when the product has been obtained, the decimal point is located as many places from the right as there are decimal places in *both* the multiplicand and the multiplier. To multiply, decimal points are not aligned.

Example Problem: Multiply 18.6 by 5.27.

Solution:

```
    1 8.6
 ×  5.2 7
 --------
   1 3 0 2
   3 7 2
  9 3 0
 --------
  9 8.0 2 2
```

Since there is one decimal place in the multiplicand and two in the multiplier, the product contains three decimal places.

DIVISION

Decimals are divided in the same way as whole numbers but with one additional step involving the location of the decimal point in the quotient. Just as in multiplication, the decimal point must be accurately located for the answer to be correct. To locate the decimal in the quotient, subtract the number of decimal places in the divisor from those in the dividend.

To work a division problem, it is best to relocate the decimal points before dividing as illustrated in the following problem.

Example Problem: Divide 0.875 by 0.5.

Solution:

```
                      1.7 5
0.5 / 0.8 7 5  =  5 / 8.7 5
```

Move each decimal point one place to the right and eliminate the zeros, making the divisor the whole number 5 and the dividend 8.75. Place the decimal point in the quotient directly above the decimal point in the dividend *before* starting to divide. This will assure getting the decimal point in the answer correctly positioned.

Example Problem: Divide 93.6404 by 3.71.

Solution:

```
3.7 1 / 9 3.6 4 0 4
```

Move the decimal point in the divisor, making it the whole number 371. Then move the decimal point in the dividend two places to the right also, making it read 9,364.04. Place the decimal point in the quotient directly above the dividend and divide. The problem would read

```
           2 5.2 4
  3 7 1 / 9,3 6 4.0 4
          7 4 2
          -----
          1 9 4 4
          1 8 5 5
          -------
            8 9 0
            7 4 2
            -----
            1 4 8 4
            1 4 8 4
            -------
```

When division problems result in remainders, they are easily handled by the use of decimals. By adding zeros to the right of the decimal point in the dividend, the process of division can be continued until no remainder is left, or it is so small that it becomes insignificant.

Example Problem: Divide 45 by 8.

Solution:

```
      5.6 2 5
  8 / 4 5.0 0 0
      4 0
      ---
        5 0
        4 8
        ---
          2 0
          1 6
          ---
            4 0
            4 0
            ---
```

If the division continues with a remainder after several decimal places have been recorded in the quotient, it may be desirable to round off the number. Rounding off an answer means to approximate the answer after a certain number of figures are known, rather than to complete the division until no remainder is left.

Example Problem: Find the quotient of 278 $\div$ 34 to the nearest hundredth.

Solution:

```
            8.1 7
  3 4 / 2 7 8.0 0 0
        2 7 2
        -----
            6 0
            3 4
            ---
            2 6 0
            2 3 8
            -----
              2 2
```

Note that after finding two decimal places in the quotient, there is a remainder of 22. To determine the answer to the nearest hundredth, the division may be carried out to get another figure in the quotient. In this case, bring down the third zero, divide 34 into 220 to get 6, which is added to the quotient to make it 8.176. Since 8.176 is closer to 8.18 than it is to 8.17, the number is rounded off to 8.18.

A second method of determining the rounded off answer involves a shortcut. If the remainder 22 is more than half of the divisor 34, the quotient is rounded up to 8.18. If the remainder had been less than the divisor, the answer would have stayed 8.17.

CONVERTING FRACTIONS

For convenience in taking measurements, it may be necessary to change decimal fractions to common fractions, which can be read on a steel scale or a rule. Likewise, some data necessary to work problems might be given in common fractions and some in decimal fractions, making it necessary to change the common fractions to decimal fractions.

Changing common fractions to decimals is called finding their decimal equivalents. Table 1.3 gives the decimal equivalents to the common fractions commonly used in industry, including those used in the machine and general repair shops. Many machinists and mechanics have memorized the equivalents they use frequently.

To change a decimal fraction to a common fraction, give the decimal its numerical denominator and then reduce to lowest terms.

Example Problem: Change 3.625 to a common fraction.

Solution:

$$3.625 = 3\tfrac{625}{1,000} = 3\tfrac{5}{8}$$

Since decimal fractions are based on 10, 100, 1,000, etc., and since scales are graduated in $\frac{1}{4}$, $\frac{1}{8}$, $\frac{1}{16}$, etc., a decimal fraction often cannot be changed to a common fraction on the scale. Usually the decimal can be changed to a common fraction very close to one on the scale. In changing a decimal fraction to an approximate common fraction, multiply the decimal by the denominator of the fraction to which it is to be converted. The product will be the numerator of the new fraction.

Example Problem: Convert 0.94 to the nearest sixteenth.

Solution:

$$0.94 \times 16 = 15.04$$

Discard the .04 and write 15 as the numerator of the new fraction, making it 15/16. Thus 0.94 ≅ 15/16. (The operational sign for *equals approximately* is ≅.)

Table 1.3
Decimal Equivalents

Fraction	Decimal Equivalent	Fraction	Decimal Equivalent
1/64	0.015625	33/64	0.515625
1/32	0.03125	17/32	0.53125
3/64	0.046875	35/64	0.546875
1/16	0.0625	9/16	0.5625
5/64	0.078125	37/64	0.578125
3/32	0.09375	19/32	0.59375
7/64	0.109375	39/64	0.609375
1/8	**0.125**	**5/8**	**0.625**
9/64	0.140625	41/64	0.640625
5/32	0.15625	21/32	0.65625
11/64	0.171875	43/64	0.671875
3/17	0.1875	11/16	0.6875
13/64	0.203125	45/64	0.703125
7/32	0.21875	23/32	0.71875
15/64	0.234375	47/64	0.734375
1/4	**0.25**	**3/4**	**0.75**
17/64	0.265625	49/64	0.765625
9/32	0.28125	25/32	0.78125
19/64	0.296875	51/64	0.796875
5/16	0.3125	13/16	0.8125
21/64	0.328125	53/64	0.828125
11/32	0.34375	27/32	0.84375
23/64	0.359375	55/64	0.859375
3/8	**0.375**	**7/8**	**0.875**
25/64	0.390625	57/64	0.890625
13/32	0.40625	29/32	0.90625
27/64	0.421875	59/64	0.921875
7/16	0.4375	15/16	0.9375
29/64	0.453125	61/64	0.953125
15/32	0.46875	31/32	0.96875
31/64	0.484375	63/64	0.984375
1/2	**0.5**	1	1

Changing a common fraction to a decimal fraction is a very simple process and consists of merely setting down the numerator and dividing it by the denominator.

Example Problem: Express 9/16 as a decimal fraction.

Solution:

```
       .5625
  16 / 9.0000
       80
       100
        96
         40
         32
          80
          80
```

PRACTICE PROBLEMS

1. Express the following numbers in words:

   a. 0.003 ________________

   b. 0.625 ________________

   c. 120.04 ________________

   d. 8.3745 ________________

2. Express the following fractions as decimal fractions and as common fractions:

   a. one hundred fifteen thousandths ________________

   b. seventy-six and seventy-six hundredths ________________

   c. five thousand and six tenths ________________

   d. three thousand, one hundred twenty-five and eight tenths ________________

3. Bill received $316.89 on pay day. He plans to pay the following bills out of his check: telephone, $24.50; charge card, $35.00; gas, $41.23. How much will he have left? ________________

4. A contractor receives \$1,543.75 for a job. His expenses on the job are as follows: labor, \$375.35; transportation, \$53.65; materials, \$544.60. How much is his profit? ______

5. Joe bought \$10 worth of gasoline at \$1.12 per gallon. How many gallons did he get? ______

6. Eighteen shallow wells are to be pumped by electric motors. A test shows that 2.36 hp is required for each well. How much power will be required for pumping all of the wells at once? ______

7. One gallon of 45°API gravity crude oil weighs 6.675 pounds. What will 8,000 gallons of such oil weigh? ______

8. A group of fifteen employees decide to donate an equal amount each for the purchase of a microwave oven for their lunch room. The price of the oven is \$329.00 plus a tax of \$16.45. How much will each person have to pay? ______

9. The depth of a groove in a wireline sheave should be 1.75 times the nominal diameter of the wireline. How deep should the groove be for a 7/8" line? ______

10. The distance across the flats (W) of a machine bolt is always equal to one and one half times the diameter of the bolt plus 1/8". Using decimals, find the dimension of W in the sketch below. ______

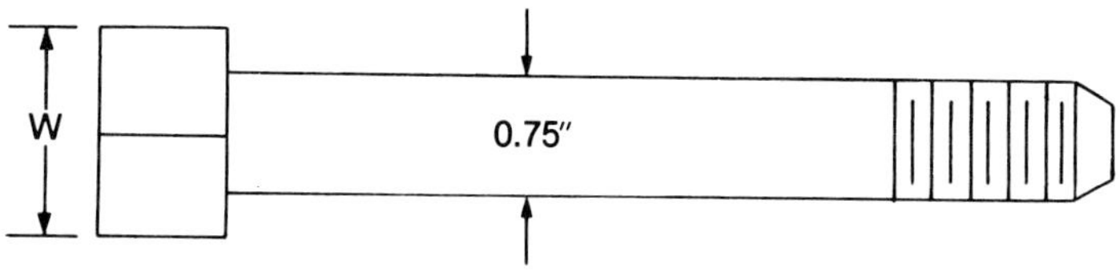

Roots and Powers

The *root,* or base, of a number or quantity is one of the equal factors which, when multiplied together, produce a given number or quantity. The number of times the root number is multiplied by itself is called the *power* and is indicated by an *exponent.* An exponent is the small number written to the right and a little above the root number or quantity.

SQUARES AND SQUARE ROOTS

The product of a number multiplied by itself is the *square* of that number. Thus, the square of 10 (10^2, read "ten squared") is 10 × 10, or 100. The *square root* of the product 100 is 10.

The common way to indicate that the square root of a quantity is to be found is to precede the quantity with a radical sign ($\sqrt{\ }$) and extend the line above all numbers to be considered when finding the square root; for example, the square root of (3 times 12) would be written:

$$\sqrt{3 \times 12}.$$

To solve this problem,

$$\sqrt{3 \times 12} = \sqrt{36} = 6.$$

Another way to indicate square root is to enclose the quantity in parentheses and use the exponent ½. Then $\sqrt{100}$ and $(100)^{1/2}$ both mean that the square root of 100 is to be found.

Many mathematical problems require extracting square roots of numbers. Notable examples are finding the radius of a circle when the area is known and finding the length of one side of a right triangle when the other sides are known, as will be seen in the chapter on geometry.

Example Problem: Find the square root of 1,806.25.

Solution:

```
     4  2. 5
   √18 06.25
    16
 80/ 2 06
  2  1 64
 82
   840/ 42 25
     5  42 25
   845
```

1. Divide the number into groups of two figures each, counting in each direction from the decimal point. There will be as many digits in the root as there are groups of numbers in the original number. In this case, there are two whole numbers and one decimal place in the root.
2. Find the largest number which, when squared, will be less than or equal to the first left-hand group. This is 4.
3. Record this number above the first group. It will be the first number in the root.
4. Square the number 4 in the root and subtract the product 16 from 18, as in long division.
5. Bring down the remainder 2 and the next group (06) for a new dividend.
6. Multiply the number 4 in the root by 20 (a trial number) and use the product 80 as a trial divisor.
7. Find the number of times the trial divisor 80 is contained in the dividend 206 and place this number 2 as the second number in the root.

8. Multiply the divisor thus obtained (82) by the second digit in the root (2) and subtract the product 164 from the dividend 206.
9. Bring down the next group of figures (25) beside this remainder 42 for a new dividend 4,225.
10. Multiply the two figures in the root (42) by 20 and use the product 840 as a trial divisor.
11. Add the number of times this trial divisor 840 is contained in the dividend 4,225 to the trial divisor, giving a divisor of 845.
12. Record the quotient 5 as the third digit in the root and multiply by the divisor 845, obtaining the product 4,225. Subtraction leaves no remainder, so the square root of 1,806.25 is 42.5.
13. To check, multiply 42.5 times 42.5.

```
    4 2.5
  × 4 2.5
  -------
  2 1 2 5
  8 5 0
1 7 0 0
---------
1,8 0 6.2 5
```

As can be seen by the example, finding the square root of a number is a laborious process. Using a calculator greatly simplifies this operation, especially if the calculator offers the square root function. However, the problems in this chapter should be solved *without* the use of a calculator.

OTHER POWERS

A number or symbol may be raised to any power. For example, 5^4 (read as "five to the fourth power") means

$$5 \times 5 \times 5 \times 5, \text{ or } 625.$$

The fourth root of 625 is 5. The square root of 625 is 25 because 25^2, or 25×25, is 625. Letters or other symbols are often used to represent the root quantity, especially in algebraic equations such as

$$a^2 + b^2 = c^2.$$

Example Problem: $23^3 + 5^5 = ?$

Solution: Solve each expression, then add together.

```
   23          5       12,167
 × 23        × 5     +  3,125
 ----        ---     --------
   69         25       15,292
  46         × 5
 ----        ---
  529        125
 × 23        × 5
 ----        ---
 1587        625
1058         × 5
-----        ---
12167       3125
```

The answer is 15,292.

A simpler way to multiply *like* base numbers with exponents is to add the exponents of the numbers and solve the final term:

$$10^5 \times 10^2 = 10^7 = 10{,}000{,}000.$$

To divide *like* base numbers with exponents, subtract the exponents:

$$85^4 \div 85^3 = 85^1 = 85.$$

Exponents and exponential functions—such as logarithms, antilogarithms, and natural logarithms—are used in the study of mathematics. The *logarithm* (commonly called *log*) of a number is the exponent of the base 10 that will make the "10" equal to that number. For example, 100 equals 10^2, making 2 the logarithm of 100 (written as log 100 = 2). Not as simple is the log of 532:

$$\log 532 = 2.7359$$

since

$$532 = 10^{2.7359}$$

Reversing the operation gives the *antilogarithm;* that is, the antilogarithm of 2.7359 is 532. The *natural logarithm* is so-called because it comes about naturally in a mathematical process. The base of this system of logarithms is designated by *e,* and its numerical value is approximately 2.71828. Despite its awkward appearance, it is a vitally important factor in many applications of higher mathematics.

Scientific calculators or logarithm tables are used to find the logs of numbers. These functions will not be used in this text.

PRACTICE PROBLEMS

1. Find the square root of 784. ________________

2. Give the quantity of 14^3. ________________

3. A square plot of ground measures 6 miles on each side. If the area of this plot equals to one of its sides squared, how many square miles are contained in the plot of ground? ________________

4. Extract the square root of 10.24. ________________

5. $62^2 + 10^4 - \sqrt{25} = ?$ ________________

SELF-TEST
1. The Number System

Multiply each question/problem answered correctly by five to arrive at your percentage of competency.

1. In 1961 the estimated world population was two billion, two hundred fifty million. In 1982 it was four billion, four hundred ninety-two million. What was the population growth in this 21-year period?

   ____

2. If an operator owned a 3/5 interest in a lease and later sold 1/8 interest, what would his remaining interest be?

   ____

3. Three metal sheets are 3/16″, 3/8″, and 9/32″ in thickness. What is their total thickness?

   ____

4. Ignoring the collar diameter, how much clearance is there between a string of 7″ casing (ID, 6¼″) and a string of seamless tubing (OD, 2⅛″)?

   ____

5. Find the outside diameter of the pipe shown in the sketch below.

   ____

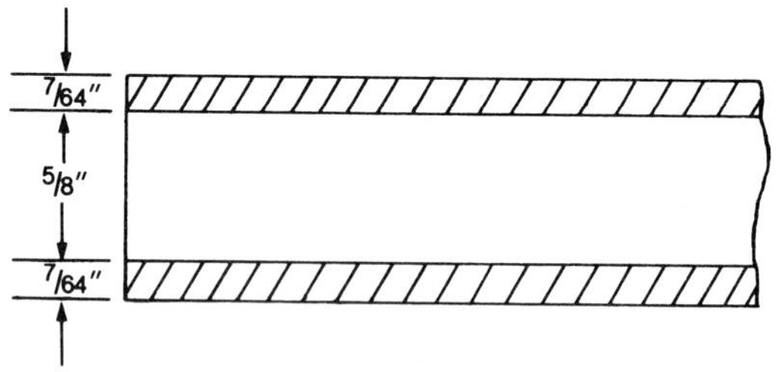

6. A lease produced 7,952 barrels of oil in a 30-day period. The oil sold for $20 a barrel. The royalty owner received 1/8 and the operator 7/8. How much money did the royalty owner receive for the period?

   ____

   The operator? ______________________

7. A simplex, double-acting pump pumps 7/8 of a gallon per stroke. At 35 strokes per minute, how many gallons are pumped per hour?

____

Per day? ______________________

8. Convert the following common fractions to four-place decimal fractions and the decimal fractions to common fractions of lowest terms:

a. 9/16 = ______________________

b. 3/32 = ______________________

c. 31/64 = ______________________

d. 2.65625 = ______________________

e. 0.0625 = ______________________

9. A trucker has to deliver a load in four days. His destination is 1,832 miles away. He travels 450.6, 520.8, and 250.3 miles in the first three days. How far does he have to travel the fourth day in order to meet his delivery date?

____

10. A pump delivers 0.635 gallons of water at each stroke, and it operates at 28 strokes per minute. If 42 gallons equals 1 barrel, how many barrels will it pump in 6 hours?

____

11. Calculate the weight of the following shipment of material:

a. 12 stiffeners that are 10′0″ long weighing 2.22 pounds per foot

____

b. 10 stiffeners that are 8′0″ long weighing 2.22 pounds per foot

____

12. The electric motor on a centrifugal circulating pump used electricity at the rate of 1.65 kilowatts. The cost of electricity for the pump was 5.33 cents per kilowatt-hour. What would be the cost of electricity to operate the pump for 24 hours?

____

13. Discounting waste, how many pins 2.75" long can be cut from a ½" bar that is 14¾' long?

____

14. The cylinders of a gasoline engine are 4¾" in diameter, and the manufacturer recommends that pistons be fitted with a clearance of 0.0035". What diameter size should the pistons for this engine be? (Express as a four-place decimal.)

____

15. A lease is sold for $380,000, one-half of which is paid in cash. One-fourth of the production is to be applied to the remainder of the note. Oil sells at $29.00 per barrel, and the lease has an allowable of 64 barrels per day. (Round off answers to the nearest whole numbers.)

a. How much is still owed on the note?

____

b. How many barrels must be sold to account for the balance owed?

____

c. How many barrels must be produced on the lease to pay off the note?

____

d. What amount of the daily revenue from the lease will be applied to the note?

____

e. How many days will it take to pay off the note?

____

16. A drilling crew adds 2¼ pounds of CMC per barrel of mud in order to treat it. How many barrels can be treated with twelve 100-pound sacks of CMC?

____

17. Solve the following problems:

a. $3^3 - 4^2 =$ ______________________

b. $\sqrt{36} \times 58 =$ ______________________

c. $92^4 =$ ______________________

d. $\sqrt{42.27 - 2} =$ ______________________

e. $10^6 \div 10^2 =$ ______________________

18. The pipe dimensions of API standard threaded line pipe (*API Spec 5L*) show the following outside and inside diameters. What are the wall thicknesses?

	Nominal Pipe Size (Inches)	Outside Diameter	Inside Diameter	Wall Thickness
a.	½	0.840	0.622	__________
b.	1	1.315	1.049	__________
c.	2	2.375	2.067	__________
d.	2½	2.875	2.469	__________
e.	3	3.500	3.068	__________

19. The recoverable butane content of a certain gas is 0.75 of a gallon per 1,000 cubic feet (0.75 gal/Mcf), the throughput of the plant is 15,000,000 cubic feet per day, and the wholesale price of the butane is 60.9 cents per gallon.

   a. How many gallons of butane will be recovered in 24 hours?

   ____

   b. How many gallons will be recovered in three weeks?

   ____

   c. How many gallons will be recovered in thirty days?

   ____

   d. What is the value of the daily production?

   ____

   e. What is the value of the annual production?

   ____

20. The following problems are designed for practice in mental arithmetic.

   a. If the daily production from a well is 150 barrels of fluid (oil and water) of which 47.9 barrels is oil, the amount of water produced in 30 days would be 1,437; 2,053; 3,063; 7,185 barrels.

   ____

b. Twenty-five pounds of red brass at 34 cents per pound would cost $40.00; $8.50; $85.00; $15.50.

____

c. The decimal equivalent of 11/4 is 2.65; 2.55; 2.75; 2.25.

____

d. The time and one-half rate for a man whose normal hourly pay is $6.18 would be $13.09; $12.27; $8.36; $9.27 per hour.

____

e. If 0.064 barrels of liquid are removed from each thousand cubic feet (Mcf) of gas at a gasoline plant, then 640; 6,400; 640,000; 64,000 barrels would be removed from 100 MMcf.

____

2 THE CALCULATOR

Objectives

Upon completion of chapter two, the student will be able to—

1. Choose a calculator that matches calculation needs with calculator design.
2. Explain the differences and likenesses of the three types of logic commonly used in calculators.
3. Describe the functions of the more commonly used keys on a calculator.
4. Solve addition, subtraction, multiplication, and division problems on a calculator.
5. Solve square root and percentage problems using a calculator.
6. Perform chain calculations using the memory function of a calculator.

The advent of the calculator marked a major technological advancement, and during the past few years, these devices have surpassed even the most hopeful predictions.

Earlier models offered the four basic functions: addition, subtraction, multiplication, and division. The functions of percentage and square root were added later. The more sophisticated models include preprogrammed scientific operations allowing rapid calculations of problems that would require many steps on the four-function calculator.

The calculator eliminates the tedious aspects of carrying through a calculation. It can prove to be a valuable tool for solving mathematical problems related to industrial operations. In drilling, if a kick occurs, a driller may use a calculator to quickly determine new mud weight, circulating pressures, and surface-to-bit strokes (Appendix A-1). Pipeline construction workers may use it to figure pipe buoyancy, while refinery employees may use it to determine heat transfer and material balance. In virtually every industry, the calculator is used for estimating and costing materials and labor for a job.

Choosing a Calculator

Calculators are available in a wide array of capabilities and prices. The least expensive models perform the four basic operations along with percentages, square roots, and a single memory. The more expensive models offer trigonometric and logarithmic functions, among others. Also, these models are equipped with a memory that can store multiple instead of single bits of information. Some of these models are programmable—the calculator can automatically perform other functions not included in the given selection.

CALCULATOR FEATURES

In choosing a calculator, it is important to match calculation needs with calculator design. Very involved calculations warrant a more sophisticated model than those calculations requiring only the four basic operations. The following are features to consider in the selection of a small desk or pocket calculator (often referred to as a handheld calculator):

1. Power supply choices
   a. AC power (AC adapter/charger)
   b. Regular penlight batteries
   c. Rechargeable penlight batteries, or battery pack, with AC adapter/charger
   d. Solar cell—powered by light
   e. Mercury or silver oxide disc cells

2. Keyboard with conventional symbols (+, −, ×, ÷, =) that permit calculations to be made in the same order in which they are normally written
3. Full-floating decimal point, a function that places the decimal point automatically and displays the maximum number of places possible
4. Constant operation, which allows a constant number to be added to, subtracted from, multiplied by, or divided into a series of variable numbers with a minimum number of keystrokes
5. Easy-to-read display—size, color, intensity, and visibility

The three principal parts of a small calculator include a *display* where lighted or liquid crystal display (LCD) symbols appear and/or a paper tape on which characters are printed; a *keyboard* that offers a selection of keys; and one or more *registers* that cannot be seen in which numbers are stored during calculations. The symbols can be found printed directly on the keys, printed on the case near the keys, or both ways. In this chapter, reference to a key will be by its symbol regardless of its printed location.

Some typical calculator keyboards are shown in figure 2.1. The calculator to be used in conjunction with this text should be able to show eight or more digits in the display and must have the following:

1. Number keys, which include the ten digits and the decimal point:

[0] [1] [2] [3] [4] [5] [6] [7] [8] [9]

Fig. 2.1. Typical calculator keyboards

2. Operation keys for the four basic arithmetic operations:

[+] [−] [×] [÷]

3. An equals [=] key and, for reverse polish (p. 43), an [ENTER] key.

Most small calculators also offer the square root [√] and percent [%] keys, which will be useful for this text. Scientific calculators should provide at least the following keys:

[y^x] [x^2] [x^y] [SIN] [COS] [TAN] [+/−] or [CHS] [LOG] [LN]

TYPES OF LOGIC

A major consideration in the selection of a calculator is the method of entry, that is, the language or logic. Basically, three major types of logic are used in calculators: arithmetic, algebraic, and reverse polish (pronounced pō-lish).

Arithmetic logic. The arithmetic logic calculator is easy to use and is identifiable by its usually simple keyboard. The operations such as addition, subtraction, multiplication, and division are done in the same order in which the keys are pressed. (Keys pressed are shown in boxes; unboxed numbers are the displayed answers.) For example, pressing the following keys in the order shown will result in the answer 35 appearing on the display (fig. 2.2):

[4] [+] [3] [×] [5] [=] 35

ENTRY	DISPLAY
4	4
+	4
3	3
×	7
5	5
=	35

Fig. 2.2. Arithmetic logic

The 4 and 3 are added together and then multiplied by the 5. Arithmetic logic is common in less expensive models of calculators and may contain, in addition to the four basic operations, percentage, square root, and memory functions.

Algebraic logic. A calculator with algebraic logic operates in the same way as one with arithmetic logic except for one important difference—it performs operations in a specific order. Powers (y^x) are done first, then multiplication (×) and division (÷), and finally addition (+) and subtraction (−). This same order is used to compute algebra problems (chapter 5). For example, to compute

$$2 + 4 \times 3^2$$

press the keys in the following order (fig. 2.3):

[2] [+] [4] [×] [3] [x^2] [=] 38

ENTRY	DISPLAY
2	2
+	2
4	4
×	4
3	3
x^2	9
=	38

Fig. 2.3. Algebraic logic

When the [=] key is pressed, the calculator begins to work, but it does not perform the operations in the order entered. First, the calculator will compute 3 to the 2nd power, producing 9. Next, it multiplies the 9 times 4 and then adds the resulting 36 to the 2 to get the answer 38.

Calculators with algebraic logic may be inexpensive, medium-priced, or expensive, depending on their capabilities. In addition to the four basic operations, percentage, and square root, many medium-priced and more expensive models can square the displayed number, calculate the inverse of the displayed number, perform operations within parentheses first so that the contents will be treated as one unit, and change the sign of a displayed number. An algebraic calculator may have a different kind of memory system. Some calculators store many pieces of information instead of just one by pressing location keys after the memory storage key. Information can be stored and recalled at any time.

Reverse polish logic. The calculator using reverse polish logic is akin to a computer and, once mastered, is a versatile and dependable tool. The reverse polish calculator works with a memory stack, which can be thought of as a ladder with several rungs. The display represents the bottom rung of the ladder, and each rung above the display is a memory level capable of holding one piece of information. When the [ENTER] key is pressed, the displayed number climbs one rung up the ladder and also pushes whatever else is on the ladder up one rung. With reverse polish logic, the numbers that are keyed in first are worked on last. When a number is entered, it is stored in the memory stack for later use.

The easiest way to understand how this process works is to go through a problem and follow the sequence of operations on the display and in the memory of a reverse polish calculator (fig. 2.4). To solve the problem

$$5 + (2 \times 4) = ?$$

the following sequence is entered into the calculator:

[5] [ENTER] [2] [ENTER] [4] [×] [+] 13.

ENTRY	DISPLAY	MEMORY
5	5	
ENT	5	5
2	2	5
ENT	2	2 5
4	4	2 5
×	8	5
+	13	

Fig. 2.4. Reverse polish logic

If the problem is stated

$$(5 + 2) \times 4 = ?$$

the calculator sequence would be

[5] [ENTER] [2] [+] [4] [×] 28.

To illustrate the differences in the arithmetic, algebraic, and reverse polish logics, review the following problem and consider the calculator's interpretation of the keystrokes when entered in this order:

$$2 + 3 \times 4 =$$

The arithmetic logic calculator will interpret the keystrokes as

$$(2 + 3) \times 4 = 20.$$

The algebraic logic calculator will perform the multiplication first and will work the problem as

$$2 + (3 \times 4) = 14.$$

The reverse polish logic calculator could not work with this sequence of keystrokes. For the arithmetic interpretation, they would have to be changed to

[2] [ENTER] [3] [+] [4] [×]

and for the algebraic interpretation to

[2] [ENTER] [3] [ENTER] [4] [×] [+].

Computer people have realized the advantages of reverse polish logic given that an expression is scanned from left to right and therefore can be executed immediately as it is encountered. Unlike the algebraic logic where the operation signs are sandwiched between the numbers, in reverse polish they are generally entered last and are not stored. Thus, reverse polish logic has the advantage over the algebraic logic in that it requires less storage space on a calculator chip.

Many companies manufacturing the small, nonscientific calculators use arithmetic logic with single memory storage. Hewlett-Packard was the first manufacturer to use reverse polish logic in scientific calculators and continues to produce them. Texas Instruments and others have opted for the algebraic logic in their scientific calculators.

Understanding Calculator Functions

In using a calculator, the two most important considerations are (1) how to enter the numbers into the calculator and (2) how to tell the calculator which operation to perform. Most important to the successful operation of any calculator is reading and following the instructions provided by the manufacturer. Knowing exactly how the calculator functions will save time and frustration as well as prevent damage to the instrument through improper use.

BASIC OPERATIONS

The following basic operating functions are found on most, if not all, small calculators in use today. The corresponding keys are shown in the margin.

Power switch. Turns the calculator on and off. It also clears the display and erases whatever is stored in the memory unless the calculator has a continuous memory. [ON/OFF]

Numbers 0-9. Displays the number pressed and keeps it for calculations.

[7] [8] [9]
[4] [5] [6]
[1] [2] [3]
[0]

Clear all. Clears all or erases whatever was in the calculator, so a new problem can be started. [C] or [CA]

Enter. Enters displayed number into calculator memory; used with reverse polish logic. [ENT]

Clear entry or clear display. Clears only what is in the display or the last entry, enabling the user to correct any mistake made during entry. This function may be combined with the clear-all function for a dual-duty key: touched once, it clears the last entry; touched again, it clears all. [CE] or [CD]

Period. Inserts a decimal point where desired. [.]

Addition. Instructs the calculator to add the next entered (or previously entered) quantity to the displayed number. [+]

Subtraction. Instructs the calculator to subtract the next (or previously entered) quantity to the displayed number. [−]

Multiplication. Multiplies the displayed number by the next entered quantity. [×]

Division. Divides the displayed number by the next entered quantity. [÷]

Equals. Completes previously entered operation and displays result; used in arithmetic and algebraic logic calculators. [=]

Percent. Shifts the decimal point of the number shown in the display two places to the left. [%]

Square root. Finds the square root of the displayed number. [√]

MEMORY FUNCTIONS

Most calculators today come with a memory. In the less expensive models, the memory has only one level, so the calculator can retain only one piece of information. Other, more expensive calculators allow the storage of many pieces of information instead of just one by pressing location keys after the memory key. Information can be stored and recalled at any time. The following are memory functions found on various calculators.

Memory addition. Adds the displayed number to that number already stored in the memory, leaving the result in the memory. [M+] or [Σ]

Memory subtraction. Subtracts the displayed number from that number already stored in the memory, leaving the result in the memory. [M−]

Memory multiplication. Multiplies the number already stored in the memory by the number shown in the display, leaving the result in the memory. [M×] or [PROD]

Memory division. Divides the number already stored in the memory by the number shown in the display, leaving the result in the memory. [M÷]

Recall from memory. Recalls memory contents and displays but does not change memory contents. [RM] or [MR]

Clear memory. Clears numbers stored in memory. [CM] or [MC]

Store in memory. Stores displayed quantity in a specific place, or level, in the memory as indicated by a two-digit number (referred to as *location* number and shown at right as *xx*). [STO] and [X] [X]

Recall from memory level. Recalls quantity stored in a specific level in the memory to the display for use in a calculation but retains the quantity in memory for further use. [RCL] and [X] [X]

Add to memory. Adds a displayed number (negative or positive) to a number in the memory in a specific level. [SUM] and [X] [X]

Multiply by number in memory. Multiplies a displayed number by a number stored in the memory in a specific level. [PRD] and [X] [X]

Exchange. Swaps the displayed number with that in the memory. [EXC] or [X=M]

Clear memory level. Clear data stored in a specific level as indicated by its two-digit location number. [CM] and [X] [X]

SCIENTIFIC FUNCTIONS

The scientific or slide-rule calculator provides many additional functions including trigonometric and logarithmic operations. Following are the most commonly used functions found on scientific calculators.

Square. Squares the displayed number (chapter 1). [x^2]

Reciprocal. Calculates and displays the reciprocal of the displayed number (chapter 7). [1/x]

x/y interchange. Changes the order within the calculator so that the number entered first or a previous result is interchanged with that entered second or a displayed result. [$x \rightleftharpoons y$]

Parentheses or brackets. Performs the operation within the parentheses first so that the contents will be treated as one unit. [(] [)] or [[] []]

Change sign. Changes the sign of the displayed number from positive to negative or vice versa. [+/−] or [CHS]

Pi. Recalls to the display the value of the constant, 3.14159265359, to the number of significant figures allowed by the accuracy of the calculator; for example, 3.1416 is five significant figures, abbreviated to 5 s.f. (Pi is used extensively in chapter 6.) [π]

Enter exponent. Allows exponential notation to cope with the display of very large or very small numbers. [EE] or [EEX]

Trigonometric functions. Allows entries of angles in all four quadrants (chapter 7). [sin] [cos] [tan]

Inverse trigonometric functions. Calculates the inverse functions of angles (chapter 7). [$\sin^{-1}$] [$\cos^{-1}$] [$\tan^{-1}$]

Logarithm. Calculates and displays the logarithm to base 10 of the displayed number (chapter 1). [log]

Antilogarithm. Calculates and displays the antilogarithm of the displayed number (chapter 1). [10^x]

Natural logarithm. Calculates and displays the natural logarithm of the displayed number (chapter 1). [ln]

Exponential function. Calculates and displays the natural antilogarithm of the displayed number. [e^x]

Shift. Accesses a second function of a key. [2nd] or [F]

Power. Raises the first number entered (y) to a power determined by the second number entered (x) (chapter 1). First number must be positive; second may be negative. [y^x]

xth root. Calculates the xth root of the first number entered (y) provided the number is not negative. [$\sqrt[x]{y}$]

Roll stack. Allows the user of a reverse polish calculator to review the contents of the working registers or to reposition numbers. [R↓] and [R↑]

Operating the Calculator

With the proper operation, the calculator provides a convenient and accurate means of solving problems—an integral part of every industry. Most calculators are easy to operate if the instructions furnished by the manufacturer are followed. Many operations may be obvious; however, some valuable features may be available but not obvious by looking at the keyboard. So, the first step in operating any calculator is to *read the instructions from the manufacturer.*

As pointed out earlier, a calculator functions according to its logic, or operating system. The simplest, electronically, is the arithmetic operating system. More involved calculations require the algebraic or reverse polish operating systems.

SOLVING PROBLEMS WITH THE ARITHMETIC CALCULATOR

On the arithmetic logic calculator (fig. 2.5), addition, subtraction, multiplication, and division of two numbers are done in logical order with the operation key pressed between entry of the numbers. For example, to add two numbers, the keys are pressed in this sequence:

[6] [+] [3] [=] 9.

The display will show 9 as soon as the [=] is pressed. Likewise, to subtract, multiply, or divide, keys are pressed in order of operation:

[6] [−] [3] [=] 3.
[6] [×] [3] [=] 18.
[6] [÷] [3] [=] 2.

In chain calculations, the sequence of operations is also the same as entered. For example, in the sequence

[6] [−] [4] [×] [5] [=] 10,

the 4 will be subtracted from the 6 first, then the resulting 2 will be multiplied by 5 to give the answer 10. A chain calculation where multiplication or division precedes addition or subtraction will also be treated in the same sequence; for example, in the problem

[4] [×] [5] [+] [9] [=] 29,

the 4 will be multiplied by the 5 to give 20, then the 9 will be added to 20 to give the answer 29.

Fig. 2.5. Arithmetic logic calculator

To calculate a chain of operations that requires algebraic logic with an arithmetic logic calculator, memory capability is helpful. Most calculators today have a memory. In the less expensive models, the memory has only one level, so the calculator can remember only one piece of information. The memory is used to solve one part of a problem, hold it, solve another part of the problem, retrieve the result of the first problem, and then perform a final operation using the results of both parts. For example, to work the problem

$$(12 + 5) \times (23 - 4),$$

the entry sequence must be as follows (fig. 2.6):

[1] [2] [+] [5] [=] [M] + [2] [3] [−] [4] [×] [MR] [=] 323.

ENTRY	DISPLAY
12	12
+	12
5	5
=	17
M +	17
23	23
−	23
4	4
×	19
MR	17
=	323

Fig. 2.6. Using memory functions with arithmetic logic

In this sequence the calculator stores the sum of the first two numbers in its memory, performs the subtraction in the next set, then multiplies this difference by the sum in its memory to get the answer 323. Note that the [=] does not need to be pressed to complete the subtraction; the [×] key activates the operation. Also, most calculators have some way of noting when a number other than zero is in the memory. Usually a small M appears on the display.

The calculator percent key [%] is useful for dividing numbers by 100 and for calculating markup and markdown percentages. Dividing a number by 100 merely moves the decimal point of that number two places to the left. Most calculators require the percentage number to be entered last. For example, to find 60% of 200, the keys would be pressed in the following order:

[2] [0] [0] [×] [6] [0] [%] 120.

In effect, 200 is being multiplied by 0.6 to give 120.

Example Problem: A businessman has borrowed $5,000. How much interest will he pay at 7.75%?

Solution: Enter as follows:

[5] [0] [0] [0] [×] [7] [.] [7] [5] [%] 387.5.

The businessman will pay $387.50 in interest.

For a markup or add-on problem, the [+] key is used instead of the [×] key.

Example Problem: You have bought a carload of tires for $34.50 each. How much would you have to sell each for in order to make a 35% profit?

Solution: To work this problem without a [%] key, you would multiply $34.50 times 0.35 and add the product to $34.50 to get $46.58. With a calculator, you would enter:

[3] [4] [.] [5] [0] [+] [3] [5] [%] 46.575.

You would have to sell each tire for $46.58 to make a 35% profit.

For a markdown or discount problem, the [−] key is used instead of the [+] key.

Example Problem: You normally sell an item for $398.99, and you want to discount the item by 16%. What will be the new selling price?

Solution:

[3] [9] [8] [.] [9] [9] [−] [1] [6] [%] 335.15

The new selling price would be $335.15.

Using the percent key greatly simplifies an operation.

Example Problem: Find the tax (5%) and the tip (15%) on a $6.93 luncheon; $1.50 of the luncheon was for a glass of wine that is not taxable, but you want to pay the tip on the total. How much money will you have to pay?

Solution: First find the cost of the lunch without the wine:

6.93 − 1.50 = 5.43.

Then, find the tax on 5.43 and add it to the total:

5.43 × 0.05 = 0.2715 + 5.43 = 5.70.

Next, add 5.70 to 1.50 and add 15% to all for the total amount:

5.70 + 1.50 = 7.20
7.20 × 0.15 = 1.08
1.08 + 7.20 = 8.28

All of these calculations can be handled by the calculator with a percent key by entering the following:

[6] [.] [9] [3] [−] [1] [.] [5] [0] [=] 5.43 [+] [5] [%] 5.7015
[+] [1] [.] [5] [0] [=] 7.2015 [+] [1] [5] [%] 8.2817.

So the total cost of the luncheon is $8.28.

Many arithmetic logic calculators also have the capability of finding the square root of a number. To find the square root of a number, enter the number into the display, then press the key with the radical sign (√). The calculator will show the answer in the display. To find the square root of 84, press the keys in the following order:

[8] [4] [√] 9.1651513.

Again, the [=] key does not have to be depressed to obtain the result.

USING THE ALGEBRAIC CALCULATOR

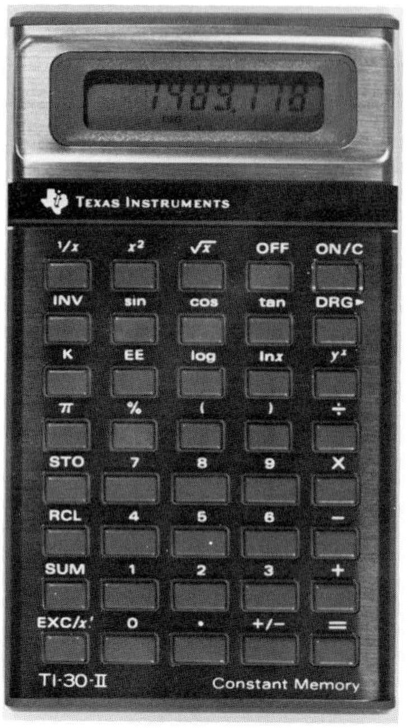

Fig. 2.7. Algebraic logic scientific calculator

In the algebraic calculator (fig. 2.7), two-number operations are performed in the same sequence as in the arithmetic calculator—numbers and operations are entered as they appear. Chain calculations are also entered as they appear; however, the sequence of operations actually taking place in the calculator follows the algebraic logic: (1) powers, (2) multiplication and division, and (3) addition and subtraction. For example, the problem

$$2 + 4 \times 3^2 =$$

would be entered as follows:

[2] [+] [4] [×] [3] [y^x] [2] [=]

When the [=] key is pressed, the calculator will begin to work, but it will not perform the operations in the order given. It will first calculate 3 to the 2nd power, producing 9; next, it will multiply the 9 by 4 to give 36; finally, it will add 36 to the 2 to give the answer, 38.

Some algebraic calculators have parenthesis keys, allowing the proper grouping of expressions. For example, the problem

$$(2 + 3) \times (4 + 5) = 45,$$

if entered without parentheses in the order shown, will be interpreted by the calculator as

$$2 + (3 \times 4) + 5 = 19.$$

Of course, 19 would be the wrong answer. By entering the parentheses as well as the numbers and operational signs in their proper order, the calculator will come up with the right answer, 45.

The algebraic calculator may have a single-level memory system or one that allows storage of more than one piece of information by pressing location keys after the memory key. To use this type of memory, the following keys can be used:

[STO], [RCL] , [SUM], and [PRD].

If the memory has many levels (storage places), the calculator must be told at what level to store the information. If the levels are designated by two-digit numbers, two digits must be entered (01, 12, etc.). To bring back a number stored in the memory, the [RCL] key and the keys for the two-digit location where the number is stored are pressed. To add a displayed number to a number in the memory, the [SUM] key and the two digit memory location keys are pressed. Likewise, to multiply a displayed number by a number in the memory, the [PRD] key and two digits are pressed. With single-level memory, the same keys are used without the location keys, or some calculators have simply the [STO] and [RCL] keys.

Some problems require both parentheses and memory functions to reach the correct answer. In this type of problem it is especially important to *clear the memory and the display before starting calculations.*

Example Problem:

$$\frac{32.64 + 18}{5.32} \times \frac{44.9 + 16.33}{81} = ?$$

Solution: The problem must be treated as two parts. The first part written for the calculator would be

(32.64 + 18) ÷ 5.32.

Enter the keys in the following order after display and memory have been cleared (fig. 2.8):

[(] [3] [2] [.] [6] [4] [+] [1] [8] [)] [÷] [5] [.] [3] [2] [=] [STO]

The answer to the first part of the problem is now stored in memory. Proceed with the second part in the same manner:

[(] [4] [4] [.] [9] [+] [1] [6] [.] [3] [3] [)] [÷] [8] [1] [=]

To multiply the two answers, press

[×] [RCL] [=]

The answer is 7.1955054, or rounded off to two places, 7.20.

ENTRY	DISPLAY
ON/C	0
(	0
32.64	32.64
+	32.64
18	18
)	50.64
÷	50.64
5.32	5.32
=	9.518797
STO	9.518797 M
C or CE	0
(	0
44.9	44.9
+	44.9
16.33	16.33
)	61.23
÷	61.23
81	81
=	0.755925
×	0.755925
RCL	9.518797
=	7.1955054

Fig. 2.8. Using memory functions with algebraic logic

When solving more complicated problems, keep in mind that a calculator with algebraic logic performs operations in a specific order. Since many calculators do not have parenthesis keys, it is necessary to solve parts of an involved problem in a specific order—not necessarily in the order of appearance. The following example problem will be solved without the use of parentheses.

Example Problem:

$$8 + \frac{30 + \sqrt{12 - (4 \times 2)}}{7 + (2 \times 3)^2} = ?$$

Solution: Solve the part of the problem in the denominator first and store it in the memory for later use. Multiply 2 times 3 so that the *answer* can be squared. If the 2 is not multiplied by the 3 first, the calculator will square the 3 and get a wrong answer. Once the problem in parentheses is solved, add the 7 to it and then store the answer, 43, in the memory. To solve for the denominator, press the keys in this order:

[2] [×] [3] [=] [x^2] [+] [7] [=] [STO]

Now solve the numerator part of the problem. The parentheses do not pose a problem here because the calculator automatically does multiplication first. Solve the part of the problem under the radical sign, then add the answer to 30:

[1] [2] [−] [4] [×] [2] [=] [√] [+] [3] [0] [=]

The answer, 32, does not need to be stored in the memory because this numerator is going to be divided by the denominator, 43, which is recalled from the memory:

[÷] [RCL] [=]

Be sure the [=] key is pressed before the [÷] key in this step. If it is not entered, the calculator will complete the division before the addition and give a wrong answer. Finally, add the 8 for the answer:

[+] [8] [=] 8.744186, or rounded off to two places, 8.74.

WORKING WITH REVERSE POLISH LOGIC

When a problem is solved using a reverse polish logic calculator (fig. 2.9), the problem must be entered in a specific order. After the parts of the problem are stored in the calculator stack, the operation keys can be pressed. Operations can be grouped into two categories. An operation with only one variable (one number to be dealt with, such as x, x^2, $1/x$) will act only on the display contents, and the answer will show in the display. An operation with two or more variables (addition, subtraction, multiplication, and division where more than one number is involved) will act on the display contents *and* on the contents of the next higher memory level in the stack. The answer will show in the display, but all stack contents will move down one level. *The numbers that are keyed in first are worked on last,* and *operation keys are entered in reverse order of their use.* Numbers are automatically stored in the memory and worked on in reverse order. After each number to be put into the memory stack is pressed, the [ENTER] key must be pressed to move the number into the stack.

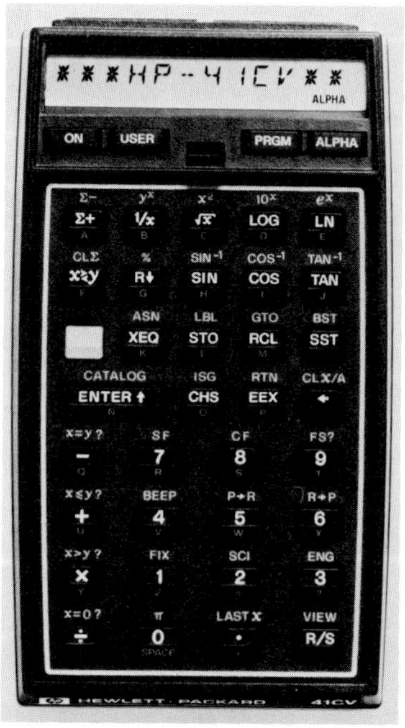

Fig. 2.9. Reverse polish logic scientific calculator ***(Photo courtesy of Hewlett-Packard Company)***

Example Problem:

$$1 + \frac{2 + \sqrt{33 - 4}}{1 + (7 \times 3)^2} = ?$$

Solution: Enter the parts of the problem in the calculator as follows:

[1] [ENTER] [2] [ENTER] [3] [3] [ENTER] [4] [−] [√] [+] [ENTER] [1] [ENTER] [7] [ENTER] [3] [×] [x^2] [+] [÷] [+] 1.0167

To understand what happens in a reverse polish calculator, follow the display and memory of the problem above as each entry is made (fig. 2.10). The number 1 was entered into the memory stack first and left there to be worked on later. Next, the numerator of the fraction was solved and stored for later use. The numerator was solved by displaying and entering the numbers from left to right. The last number, 4, was displayed but not entered so that it could be subtracted from 33 to get an answer of 29. Working backwards, the square root of 29 was found and added to the 2. The answer to the numerator, 7.38, was entered in the memory stack.

Next, the denominator of the fraction was solved in the same way, entering the numbers from left to right and then working backwards with the operations between these numbers. Because the 3 and 7 were in parentheses, they were one unit, and they had to be multiplied together first to get an answer of 21. Then the square of 21 was found and added to the 1. The result was 442, the answer to the denominator. This answer was not entered into the memory stack, as the next step was to divide the numerator by the denominator (7.38 ÷ 442). Finally, the result of the division was added to the whole number 1 to obtain the solution, 1.0167.

ENTRY	DISPLAY	MEMORY
1	1	
ENT	1	1
2	2	1
ENT	2	1 2
33	33	1 2
ENT	33	1 2 33
4	4	1 2 33
−	29	1 2
√	5.38	1 2
+	7.38	1
ENT	7.38	7.38
1	1	1 7.38
ENT	1	1 7.38 1
7	7	1 7.38 1
ENT	7	1 7.38 1 7
3	3	7 1 7.38 1
×	21	1 7.38 1
x^2	441	1 7.38 1
+	442	7.38 1
÷	0.0167	1
+	1.0167	

Fig. 2.10. Stack memory of reverse polish logic calculator

PRACTICE PROBLEMS

The following problems can be solved by any of the calculator types discussed. Experiment with your calculator, read your calculator instructions, and discover how its logic works. Then solve these problems, rounding off the answers to two decimal places.

1. $32.4 + 68.3 + 17.1 =$ __________
2. $981.503 + 576 + 834.334 - 1{,}000 =$ __________
3. $5{,}621 - 5.86 - 7.90 - 3 - 569 =$ __________
4. $53.85 \times 28 + 40.75 =$ __________
5. $(91 - 52) \times 48 =$ __________
6. $8.9 \times 0.87 \times 0.065 =$ __________
7. $(575 \times 286) \div 300 =$ __________
8. $2{,}500 \times (25 \div 50)^2 =$ __________
9. $350 + \sqrt{3{,}368 - 4} =$ __________
10. 46% of 3,500,152 = __________
11. $72 - [(530 - 468) \div (95 \times 0.052)] =$ __________
12. How far can a car be expected to travel without a refill if it gets 22 miles per gallon and its tank holds 14 gallons? __________
13. If 9.8 pounds of peaches cost $12.65, what was the selling price per pound? __________
14. A bag of nails selling for $3.50 at regular price was advertised at 20% off. If you bought 5 bags at the sale price and paid a 5% sales tax, what would be your total bill? __________
15. $\dfrac{14.9 \times (13.5 - 12.2)}{35.5 - 13.5} \times 1{,}450 =$ __________

SELF-TEST
2. The Calculator

Multiply each correct answer by five to arrive at your percentage of competency.

Solve the following problems using a calculator. Show all places.

1. Find the sum of 8, 9, 7, 5, 4, 6, 2, 1, and 3. ______________________

2. How much is 30.1 plus 3.01 plus 0.301 plus 0.0301?

   ____

3. Add 4.3, 6.75, 8.67, 12.0, and 0.16. ______________________

4. What is 10 less 1.67? ______________________

5. $6 + 5 - 14 + 4.5 - 0.5 =$ ______________________

6. $734 - 30.6 + 90.3 - 428.1 =$ ______________________

7. $(5{,}200 \times 3) - 1{,}300 =$ ______________________

8. \$1.00 + 36¢ − \$0.21 − 15¢ + \$1.00 = ______________________

9. $0.052 \times 11.2 \times 4{,}000 =$ ______________________

10. $2{,}329.6 \div 4{,}000 \div 0.52 =$ ______________________

11. $13.2 + \dfrac{290}{0.052 \times 5{,}500} =$ ______________________

12. $(12.3 - 9.5) \times 3{,}000 \times 0.052 =$ ______________________

13. $(3{,}090 \div 1.32) - (9.5 - 9.3) + 3{,}000 \times 0.052 =$

    ____

14. $6.10^3 \div (4.38 + 3.25)^2 =$ ______________________

15. $0.624 + \sqrt{3{,}481} - (600 - 490) \div 240 =$ ______________________

16. $1{,}000 + 4.5^2 \times (0.8 \times 680) =$ ______________________

17. $72 + \sqrt{530 - 468} - (95 \times 0.052) =$ ______________________

18. 12.45 + 15% = ________________

19. 6% of $1,195 = ________________

20. 3,454 less 33.3% = ________________

3 NUMBER RELATIONS

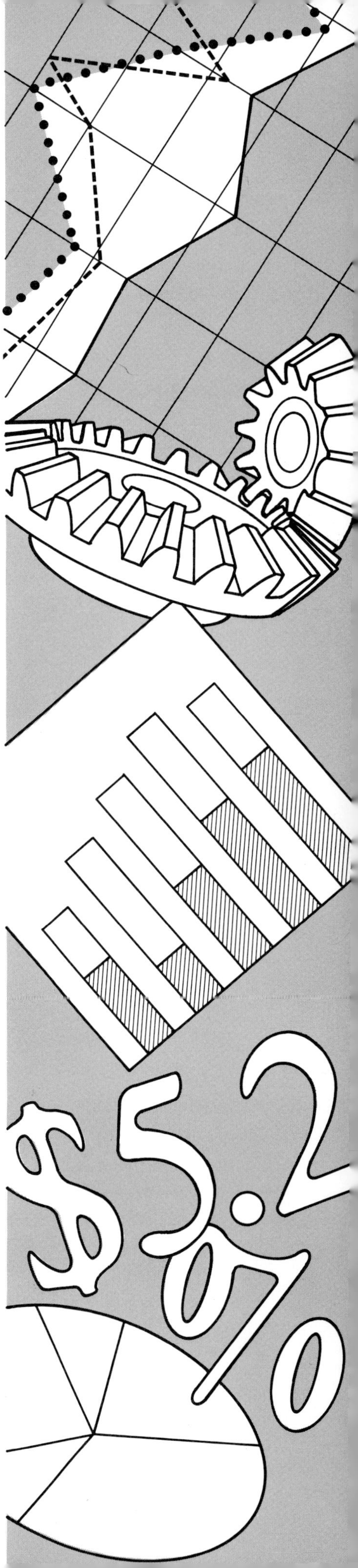

Objectives

Upon completion of chapter three, the student will be able to—

1. Discuss the relation of percent to the whole, and calculate a given percent of a given number.
2. Change percent to hundredths and hundredths to percent.
3. Solve for base, rate, or percentage in percent problems.
4. Read, write, and determine ratios of one quantity to another.
5. Solve problems involving direct proportion.
6. Solve problems involving inverse proportion, including pulley and gear ratio problems.
7. Find the average, or mean, of a set of statistics.
8. Find the median and mode in a body of data.
9. Use reference tables for extracting information.
10. Interpolate additional numerical data from information given in a table.
11. Name the parts of a table and construct a table from given data.
12. Extract approximate statistics from a graph, noting trends.
13. Determine the best method for depicting numerical information.
14. Plot and draw a bar graph, a line graph, and a circle graph.

A relationship among numbers or a chart or graph depicting numbers can show a trend, result in a meaningful deduction, or aid in making a decision based on numerical facts. Common relationships are percentage, ratio, proportion, average, mean, median, and mode. Graphs, tables, and charts are often used to depict some of these and other relationships.

Percentage

The word *percent* denotes hundredths. Percent means for or out of each hundred. The symbol for percent is % and is used to express quantity with relation to a whole. It is only used specifically and always with a number or numeral. For example:

$$26 \text{ percent} = 26/100 = 26\% = 0.26.$$
$$14\tfrac{1}{2} \text{ percent} = 145/1{,}000 = 14.5\% = 0.145.$$

The operation of changing from percent to hundredths involves shifting the decimal point two places to the left and dropping the percent symbol. For example, to change 32% to hundredths, simply drop the % symbol and move the decimal point two places to the left, yielding 0.32. In order to change a decimal fraction to percent, move the decimal point two places to the right and add the percent symbol; for example, 0.64 becomes 64%.

Example Problem: Find 14% of 430.

Solution:

$14\% = 0.14$
$0.14 \times 430 = 60.2.$
Thus, 14% of 430 is 60.2.

Using a calculator with a % function greatly simplifies this operation to

$$430 \times 14\% = 60.2.$$

Example Problem: Find 4½% of 85.

Solution:
$0.045 \times 85 = 3.825,$

or using the calculator,
$85 \times 4.5\% = 3.825.$

BASE, RATE, AND PERCENTAGE

The terms *base, rate,* and *percentage* are often used in percentage problems. *Base* is a quantity of which a certain percentage is desired, *rate* is a desired percentage of the base, and *percentage* is the product of the rate times the base (fig. 3.1). The relationship of base, rate, and percentage can be expressed as

RATE × BASE = PERCENTAGE
↓ ↓ ↓
6% of 300 is 18

Fig. 3.1. Percent relations

$$percentage = rate \times base.$$

The three common types of percentage problems involve finding one of these elements when the other two are known. When the base and rate are known, percentage can be found by multiplying the base times the rate.

Example Problem: How much is 8% of $625?

Solution:

$625 × 0.08 = $50.00.

Example Problem: If a man earns $120 and saves 12½% of it, what is the percentage saved?

Solution:

12½% = 12.5% = 0.125
$120 × 0.125 = $15.00.
The percentage saved is $15.00.

To solve for rate with the base and percentage known, the formula is changed to

$$rate = percentage \div base.$$

Example Problem: What percent of $500 is $125?

Solution:

125 ÷ 500, or 125/500 = 0.25, or 25%.

Example Problem: An oil tank contains 130 barrels, of which 2.6 barrels are sediment and water (S&W). What percent of S&W does it contain?

Solution:

2.6/130 = 0.02, or 2% S&W.

To solve for base when the rate and percentage are known, the relationship may be expressed as

$$base = percentage \div rate.$$

Example Problem: An oilwell produces 95% water, and the allowable oil production for a specified period is 325 barrels. What is the total fluid production needed to produce this much oil?

Solution: Oil comprises 5% (rate) of the fluid:

100% − 95% = 5%.

The allowable, 325 barrels, is the percentage; and the total fluid production is the base. Then

base = 325/0.05 = 6,500 barrels.

Example Problem: 43.75 is 5% of what number?

Solution:

43.75/0.05 = 875.

PRACTICE PROBLEMS

Round off all answers to two places.

1. An item costs $52 plus 5% sales tax. What is its total price?
2. How much is 8½% of 200?
3. Six percent of a man's earnings goes into his retirement fund. If he earns $1,940 per month, what amount goes toward his retirement each month?
4. An alloy contains 83.3% tin, 5.6% copper, and 11.1% antimony. How much of each metal is contained in 500 pounds of this alloy?

   a. Tin

   b. Copper

   c. Antimony
5. In a shipment of pressure gauges, 15% were damaged, leaving 170 that could be used. How many gauges were in the shipment?
6. A supply house gives a discount of ⅕ from the prices listed in its catalog. What will be the net price of four 3-inch gate valves that list at $92 each?
7. A carpenter orders 240 two-by-fours but sends 36 of them back because they are defective. What percent did he return?

8. A job in a machine shop required 27 hours to complete. Of this, 14½ hours were spent on the lathe, 2¼ hours on the drill press, 6¾ hours on the shaper, and 3½ hours in welding. What percent of the time did each process require?

   a. Lathe __________

   b. Drill press __________

   c. Shaper __________

   d. Welding __________

9. A certain oil yields 39% gasoline. How many barrels of this crude will it take to produce 500 barrels of gasoline? __________

10. The following problems are designed to provide practice in mental arithmetic. Pencils should be used only to record answers:

    a. Find 25% of 88. (Hint: Multiplying by 0.25 is the same as dividing by 4.) __________

    b. If a man earning $400 per week receives a 25% raise and then a 20% cut in salary, what is his new salary? __________

    c. If 42 gallons of oil are removed from a full 100-barrel tank, what percent of oil is left in the tank? __________

    d. If ⅛ of a lease has been surveyed, what percent remains to be surveyed? __________

    e. What percent of 110 is 55? __________

Ratio and Proportion

RATIO

The word *ratio* means relationship and describes how two different groups of things relate numerically to one another. Saying "the ratio of trucks to tank cars is two to one" means that the number of trucks is twice as great as the number of tank cars. The ratio of one quantity to another is obtained by dividing the first quantity by the second. The term *ratio* may be better defined as the relative size of two quantities expressed as the quotient of one divided by the other. Thus, the ratio of \$6 to \$2 is said to be "three to one," because \$6/\$2 equals 3/1.

In expressing the ratio of one quantity to another, the colon may be used between the two quantities (\$6:\$2), or the ratio may be written as a fraction (\$6/\$2).

Example Problem: What is the ratio of 12 to 4?

Solution: Applying the definition for ratio, divide:

$$12/4 = 3.$$

The ratio is 3 to 1.

DIRECT PROPORTION

A proportion is a relation of equality between two ratios. The ratio of \$2 to \$4 is the same as the ratio of \$6 to \$12. Proportions are used most often when one ratio and only part of another ratio are known. Since ratios in a proportion are equal in value, the missing part of a ratio can be found. As with ratios, two common notations are used for expressing a proportion—with a colon or as a fraction:

$$2:4 = 6:12$$

or

$$2/4 = 6/12.$$

Written in either form, this proportion is read as "two is to four as six is to twelve."

The first and fourth terms of a proportion (in this case 2 and 12) are called the *extremes,* and the second and third terms (4 and 6) are called the *means.* A key consideration is that in any proportion, *the product of the extremes is equal to the product of the means.* For example, in the proportion 2:4 = 6:12,

$$2 \times 12 \text{ (extremes)} = 24$$
$$4 \times 6 \text{ (means)} = 24$$
$$24 = 24.$$

In solving problems of proportion, three terms must be known, and the fourth term can be determined by applying this "rule of three." Direct proportion involves a proportion in which each term increases or diminishes as the term on which it depends increases or diminishes (fig. 3.2).

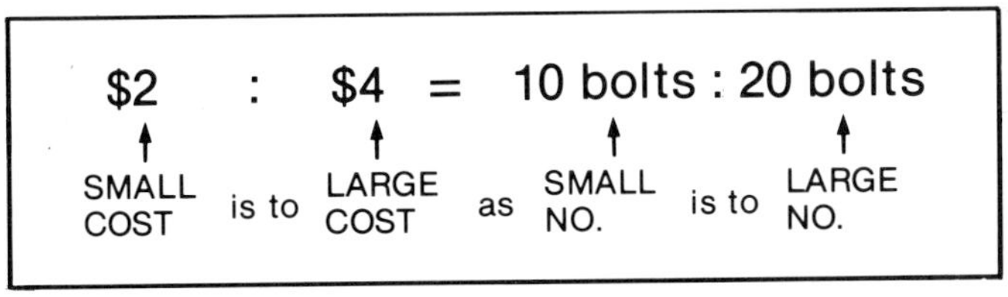

Fig. 3.2. Direct proportion

Example Problem: If 12 hacksaws cost $155.88, how much will 100 hacksaws cost?

Solution: Here there are three quantities: a small number of hacksaws, a small cost, and a large number of hacksaws. Since there is no number to express the cost of the large number of hacksaws, call the unknown quantity x and set up the proportion:

$12{:}100 = \$155.88{:}x.$

Applying the rule:

$12 \times x = 100 \times \$155.88$

$12x = \$15{,}588.$

$x = \$15{,}588 \div 12 = \$1{,}299.$*

Then 100 hacksaws would cost $1,299.

*Note: See transposition of formulas in chapter 5 to understand this calculation.

INVERSE PROPORTION

An inverse proportion, also known as an indirect proportion, is one in which each term increases as the term on which it depends decreases, or decreases as the term on which it depends increases (fig. 3.3). The ratios here vary in the opposite order. For example, the speed of a car is inversely proportional to the time needed to travel a certain distance. A car moving at 25 miles per hour can cover 35 miles in 84 minutes, while at 50 miles per hour it can go the same distance in 42 minutes.

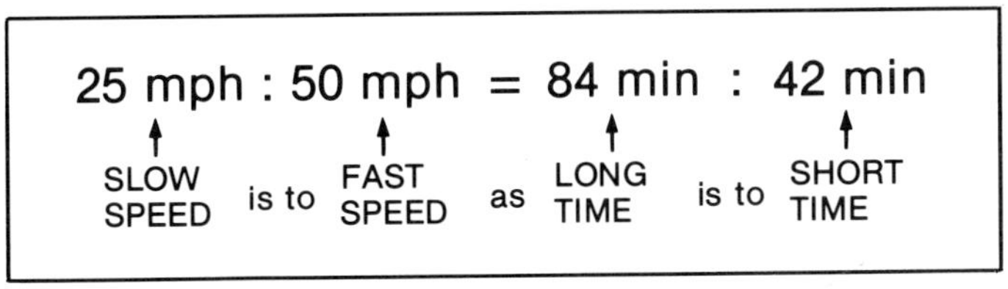

Fig. 3.3. Inverse proportion

Another example of inverse proportion involves pulleys. The bigger the pulley, the more slowly it moves. Pully B in figure 3.4 moves more slowly than the smaller pulley A. Increasing the size of pulley A will decrease its speed; likewise, decreasing the size of pulley B will increase its speed.

Fig. 3.4

Example Problem: If the speed of pulley B in figure 3.4 is 250 revolutions per minute (rpm), and it is 12 inches in diameter, how many rpm will pulley A turn if it is 8 inches in diameter?

Solution: Since the speeds of the two pulleys are inversely proportional to their diameters, write the proportion:

$$S_A : S_B = d_B : d_A$$

where

S_A = speed of pulley A
S_B = speed of pulley B
d_B = diameter of pulley B
d_A = diameter of pulley A.

Then substitute the known values and let x symbolize the speed of pulley A:

$$x:250 = 12:8$$
$$8x = 250 \times 12 = 3{,}000$$
$$x = 3{,}000 \div 8 = 375.$$

Thus, the speed of pulley A is 375 rpm.

A simple rule to remember in calculating speeds and diameters of pulleys is as follows: *The speed of the driver multiplied by its diameter is always equal to the speed of the driven pulley multiplied by its diameter.* The unknown or missing quantity is designated by some symbol such as x, and its value is found as previously shown.

This rule applied to gears would read: *The speed of the driver multiplied by its number of teeth is always equal to the speed of the driven gear multiplied by its number of teeth.*

Example Problem: When gear A in figure 3.5 makes 5 revolutions, how many revolutions will gear B make?

Solution: Using the gear ratio rule, set up the proportion:

$$S_A \times T_A = S_B \times T_B$$

where

S_A = speed of gear A
T_A = number of teeth in gear A
S_B = speed of gear B
T_B = number of teeth in gear B.

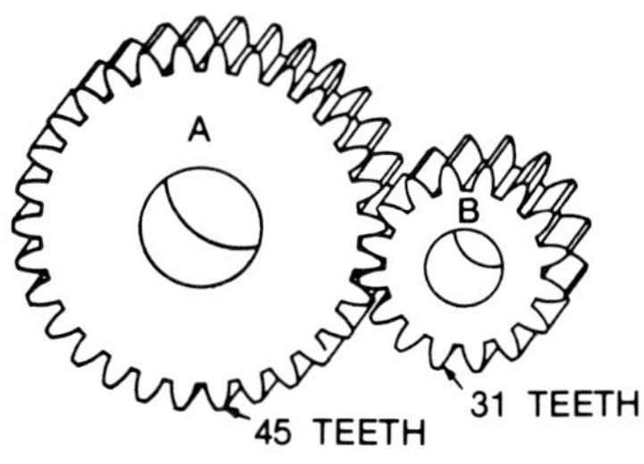

Fig. 3.5

Substituting the known values and letting x equal the unknown:

$$5 \times 45 = x \times 31$$
$$31x = 225$$
$$x = 7.258.$$

Thus, gear B will turn 7.258 revolutions each time gear A makes 5 revolutions.

PRACTICE PROBLEMS:

1. What is the ratio of 125 to 5? 25:1

2. Jones' salary is $250 and Smith's is $150. What is the ratio of Jones' salary to Smith's? 5:3

3. Soft pine wood weighs 35 pounds per cubic foot; steel weighs 490 pounds per cubic foot. Find the ratio of these densities. 14:1

4. In treating drilling mud, 60 sacks of barite are added to 100 barrels of mud to increase its weight 1 pound per gallon. How many sacks are required to raise the weight of 850 barrels of mud from 9.4 to 11 pounds per gallon? 816 sacks

5. If 3 inches on a blueprint represents 12 feet, how many feet does a line that is 10 inches long represent? 40 ft

6. A job must be completed in 6 days. The records show that a similar job was done by 20 men in 15 days. How many men must be put on the job to ensure its completion in the given time? 50 men

7. A car traveling at 48 miles per hour takes 50 minutes to go from lease A to lease B. At what rate must it run to make the trip in 30 minutes? 80 mph

8. The motor shown in the sketch below runs at 1,740 rpm. How many revolutions per minute will the pulley make? 435 rpm

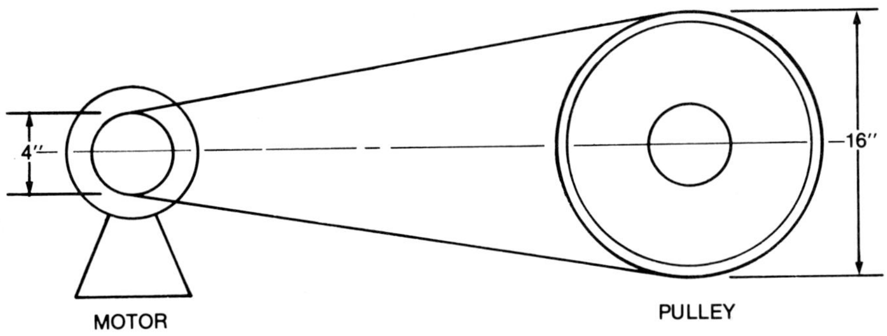

9. The motor shown in the sketch below runs at 1,800 rpm and has a 4-inch pulley. The countershaft pulleys are 8 inches and 18 inches in diameter, and the lineshaft is 16 inches in diameter. How many revolutions per minute does the lineshaft make? ____________

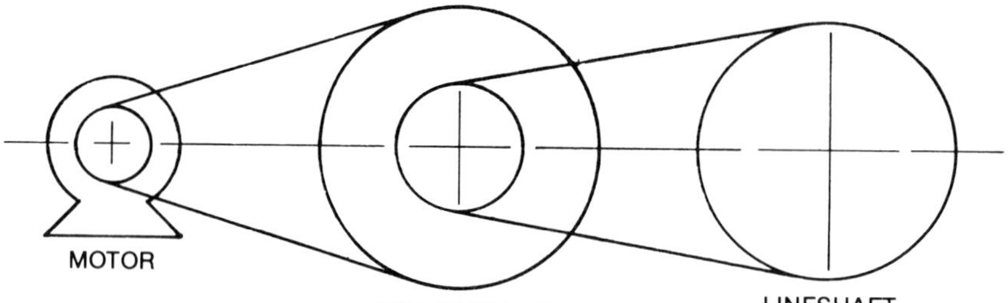

10. If a 48-tooth gear meshing with a 64-tooth gear has a speed of 165 rpm, what is the speed of the 64-tooth gear? ____________

Mean (Average), Median, and Mode

Four other terms often included in the discussion of data are *average, mean, median,* and *mode. Average* and *mean* are identical statistically; however, the term *mean* is usually used when discussions also involve *median* and *mode.*

AVERAGE, OR MEAN

The term *average,* or *mean,* may be defined as a number that typifies a set of numbers or represents the midpoint between two extremes. Certain average quantities, such as the average specific gravity of a fluid, the average cost of an item, and the average temperature of a liquid, are often desirable constants. To find the *straight* average, total the quantities and divide by the number of quantities involved.

Example Problem: Find the average cost of a handheld calculator if four telephone bid requests resulted in \$15.00, \$16.50, \$12.99, and \$11.65.

Solution: Total the costs:

$$15.00 + 16.50 + 12.99 + 11.65 = 56.14.$$

Then divide this total by the number of bid requests:

$$56.14 \div 4 = 14.035.$$

Thus, the average cost of a handheld calculator is \$14.04.

In many cases, a *weighted* average best expresses the average value. For example, a person buys 3 screws at 50¢ each and 17 screws at 80¢ each. The average cost might seem to be 65¢:

$$(50 + 80) \div 2 = 65.$$

However, more screws were bought at the higher price, so the 80¢ should be weighted more in deciding on an average cost. The simplest way to determine the true average cost is to find the total cost:

$$(0.50 \times 3) + (0.80 \times 17) = \$15.10,$$

and then divide by the total number of screws:

$$\$15.10 \div 20 = 75.5¢.$$

The average cost per screw is 75.5¢.

A simple equation for finding a weighted average is

$$A = \frac{(N_a \times V_a) + (N_b \times V_b) + (N_c \times V_c) + \ldots}{a + b + c + \ldots}$$

where

A = weighted average
N = number of items
V = value of each item
a, b, c = items.

Example Problem: A service station owner totaled his sales for the day as follows: 1,644 gallons of regular at 99.9¢/gal; 15,206 gallons of regular unleaded at $1.169/gal; and 7,310 gallons of super unleaded at $1.269/gal. What was his average earnings per gallon for the day?

Solution: Using the equation for finding a weighted average, substitute the figures:

$$A = \frac{(1{,}644 \times 0.999) + (15{,}206 \times 1.169) + (7{,}310 \times 1.269)}{1{,}644 + 15{,}206 + 7{,}310}$$

$$= (1{,}642.356 + 17{,}775.814 + 9{,}276.39) \div 24{,}160$$
$$= 28{,}694.56 \div 24{,}160 = 1.1876.$$

Thus, the average earnings per gallon for that day was $1.188.

MEDIAN

The median in a body of data is the middle value; that is, there are equally as many items of data greater than the median as there are less than the median. For example, in the following series of temperature readings,

50°, 53°, 57°, 61°, 64°, 70°, and 73°,

the median is 61° because three values (50°, 53°, and 57°) are less than 61°, and three values (64°, 70°, and 73°) are greater than 61°.

If there is an even number of values, the median may fall between two values. In the series of numbers

3, 7, 9, 11, 13, and 16,

the median, which lies between 9 and 11, is 10.

Example Problem: Find the median in the following set of speeds:

25, 30, 45, 55, 58, 60, 65, and 70 mph.

Solution: The median, which lies between 55 and 58, is 56.5 mph. To compute, add the two numbers and divide the sum by 2:

$55 + 58 = 113 \div 2 = 56.5.$

MODE

In a set of data, the mode is the value that occurs most frequently. For example, in a series of measurements such as

4″, 5″, 5″, 7″, 6″, 3″, 5″, 3″, and 5″,

the mode is 5″ because that value occurs more often than any other value in the data.

Example Problem: Find the mode in the following set of numbers:

30, 45, 65, 10, 45, 30, 50, 55, 65, 45.

Solution: The mode is 45.

PRACTICE PROBLEMS

1. In 1979, there were 2,177 rotary rigs in use. In 1980, there were 2,909 rigs, and in 1981, 3,970 rigs. What was the average number of rotary rigs operating during the three years? 3020 rigs

2. A company needed three reports duplicated. The cost of duplicating the reports was quoted as follows:
   Report A—500 copies at 35¢ per copy;
   Report B—250 copies at 50¢ per copy;
   Report C—100 copies at 70¢ per copy.
   What was the weighted average cost per copy? 370/850 = 0.44

3. Give the mean of the following numbers:
   9, 14, 52, 66, 99.9, 34.5, 7½, 47.8, and 37.5. 366.? / 9 = 40.?

4. Find the median of the following monthly wages paid by one company to its supervisors:
   $1,550; $2,350; $1,245; $2,050; $1,860. $1,860

$1,245; $1,550; $1,860; $2,050; $2,350

5. What is the mode in the following set of measurements: 15″, 1′, 30.48 cm, 18″, 1½′, 0.610 m, and 12″? ____________

Tables and Graphs

Representing numerical information in such a way that it may be clearly understood is important to anyone using numbers on the job. A random list of numbers may hide important trends, whereas some sort of table or graph based on these numbers might show these trends clearly. For example, hourly pressure readings on a gas pipeline might be presented randomly as "At 1:00 p.m., the pressure reading was 675 psig; at 2:00 p.m., 670 psig; at 3:00 p.m., 665 psig; at 4:00 p.m., 655 psig; and at 5:00 p.m., 650 psig."

The same information can be presented in tabular form (fig. 3.6) or graphically (fig. 3.7). The table shows on careful reading that the pressure is falling steadily, but the graph shows this trend at a glance.

TABLE — GAS PIPELINE PRESSURE READINGS

Time	Pressure (psig)
1 P.M.	675
2 P.M.	670
3 P.M.	665
4 P.M.	655
5 P.M.	650

Fig. 3.6. Simple table

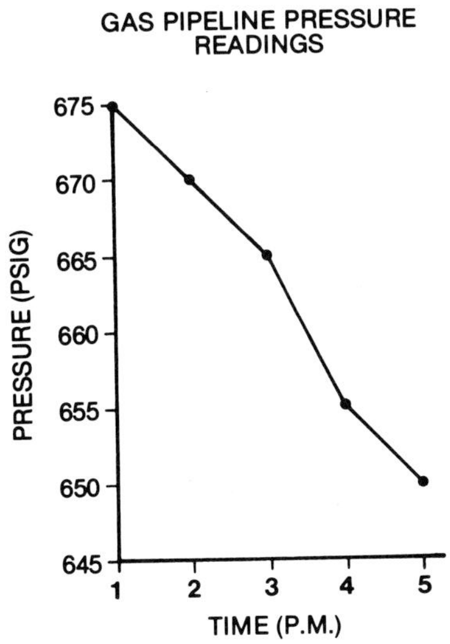

Fig. 3.7. Simple graph

USING REFERENCE TABLES

Large amounts of detailed information can be presented in a small amount of space when presented in a table. A tabular presentation is often the only way to give extensive data in an understandable way.

Information to be used for reference is often most conveniently presented in tabular form. Knowing how to read and use tables can be important when solving mathematical problems. For example, the table in figure 3.8 gives the squares, cubes, square roots, and cube roots for numbers from 1 to 50. Such a table can be used for finding any of these properties for a given number, or for finding the number if one of the properties is known.

Example Problem: Find the cube of 47, using the table in figure 3.8.

Solution: Find the number 47 in the left-hand column labeled "No." Then move horizontally across the table to the "Cube" column. The answer—103,823—will be found where the two columns intersect.

The table also may be used to find an unknown number such as a in the equation

$$a^3 = 39{,}304.$$

Locate the number 39,304 in the "Cube" column, then read the answer 34 across from it in the "No." column.

SQUARES, CUBES, SQUARE ROOTS, AND CUBE ROOTS OF NUMBERS

No.	Square	Cube	Sq. Root	Cube Root
1	1	1	1.00	1.00
2	4	8	1.41	1.26
3	9	27	1.73	1.44
4	16	64	2.00	1.59
5	25	125	2.24	1.71
6	36	216	2.45	1.82
7	49	343	2.65	1.91
8	64	512	2.83	2.00
9	81	729	3.00	2.08
10	100	1,000	3.16	2.15
11	121	1,331	3.32	2.22
12	144	1,728	3.46	2.29
13	169	2,197	3.61	2.35
14	196	2,744	3.74	2.41
15	225	3,375	3.87	2.47
16	256	4,096	4.00	2.52
17	289	4,913	4.12	2.57
18	324	5,832	4.24	2.62
19	361	6,859	4.36	2.67
20	400	8,000	4.47	2.71
21	441	9,261	4.58	2.76
22	484	10,648	4.69	2.80
23	529	12,167	4.80	2.84
24	576	13,824	4.90	2.88
25	625	15,625	5.00	2.92
26	676	17,576	5.10	2.96
27	729	19,683	5.20	3.00
28	784	21,952	5.29	3.04
29	841	24,389	5.38	3.07
30	900	27,000	5.48	3.11
31	961	29,791	5.57	3.14
32	1,024	32,768	5.66	3.17
33	1,089	35,937	5.74	3.21
34	1,156	39,304	5.83	3.24
35	1,225	42,875	5.92	3.27
36	1,296	46,656	6.00	3.30
37	1,369	50,653	6.08	3.33
38	1,444	54,872	6.16	3.36
39	1,521	59,319	6.24	3.39
40	1,600	64,000	6.32	3.42
41	1,681	68,921	6.40	3.45
42	1,764	74,088	6.48	3.48
43	1,849	79,507	6.56	3.50
44	1,936	85,184	6.63	3.53
45	2,025	91,125	6.71	3.56
46	2,116	97,336	6.78	3.58
47	2,209	103,823	6.86	3.61
48	2.304	110,592	6.93	3.63
49	2,401	117,649	7.00	3.66
50	2,500	125,000	7.07	3.68

Fig. 3.8. Reference table

INTERPOLATION IN TABLES

Even a very complete table often does not show the specific information needed in working problems. When additional figures are needed between table entries, they can be found by interpolation. To interpolate means to insert or estimate values between two known values. For example, find the square root of 800 using figure 3.8. This information is not given in the square root table, but it can be found by interpolation. Note that the table shows the square root of 784 to be 28 and the square root of 841 to be 29. The square root of 800 is between 28 and 29. To interpolate this information, find the difference in the two numbers given:

$$841 - 784 = 57.$$

Then find the difference in the number desired and the smaller number known:

$$800 - 784 = 16.$$

Thus the square root of 800 must be 16/57, or 0.28, more than the square root of 784:

$$28 + 0.28 = 28.28.$$

CONSTRUCTING TABLES

Most tables are constructed to present numerical or statistical information in the form of percentages, amount of money, number of occurrences, and so forth. In any one table, one set of information represents the primary focus of the table, and a second set of data represents varying conditions or other related variables.

A simple table is constructed with the primary set of variables (known as *boxheads*) across the top of the table and the second set of variables (known as the *stub*) down the left side of the table (fig. 3.9). The *body* of the table presents the desired statistics. The wording of the *table title* is very important and affects the amount of information that must be presented in the boxheads. Boxheads or stubs also need units of measure to qualify them, such as percent, dollars, inches, metres, degrees Fahrenheit, and so forth.

TITLE

Natural Gas Pipelines and Customers by State

BOXHEADS — STUB — BODY

State	Pipelines (miles)	Customers (thousands)
Ala.	5,301	644.4
Alaska	115	30.4
Ariz.	5,041	561.0
Cal.	7,813	6,632.2
Col.	6,219	726.6
Conn.	538	390.2
Del.	227	82.4
Total	252,012	44,941.5

Fig. 3.9. Parts of a table

Example Problem: According to projections published by the *Pipeline & Gas Journal* in 1980, the number of miles of main gas distribution utility piping to be installed between 1985 and 1990 were as follows: 1985—14,000; 1986—14,200; 1987—14,210; 1988—14,200; and 1989—15,100. The number of miles of service lines during the same period was projected to be the following: 1985—15,000; 1986—15,600; 1987—15,500; 1988—15,900; and 1989—15,900. Prepare a table that presents these statistics as well as the total number of miles of piping installed for each year.

Solution: Pick out the primary focal point of the table and use these points as boxheads: *Year, Main, Service,* and *Total*. For these heads to be meaningful, a unit of measurement (miles) should be added to Main, Service, and Total. Then, the years become the stub, and the statistics become the body as shown in figure 3.10.

PROJECTED GAS UTILITY PIPING TO BE INSTALLED

Year	Main (mi)	Service (mi)	Total (mi)
1985	14,000	15,000	29,000
1986	14,200	15,600	29,800
1987	14,210	15,500	29,710
1988	14,200	15,900	30,000
1989	15,100	15,900	31,000

SOURCE: "Gas Distribution Utility Piping Market Statistics," *Pipeline & Gas Journal,* 1980.

Fig. 3.10

PLOTTING GRAPHS

A graph is a diagrammatic representation of changes in a variable quantity in comparison with those of other variables. The prime requirement for a graph is a pair of reference axes (fig. 3.11). Axes are commonly drawn perpendicular to each other. The horizontal axis is called the *abscissa,* or X axis; the vertical axis is called the *ordinate,* or Y axis.

The location of any point on the graph may be given as its perpendicular distance from the two axes. For example, point *A* in figure 3.11 is located 5 units from the X axis and 4 units from the Y axis, or 4 across and 5 up. These dimensions are often referred to as coordinates; in this case, the coordinates of point *A* are 4, 5.

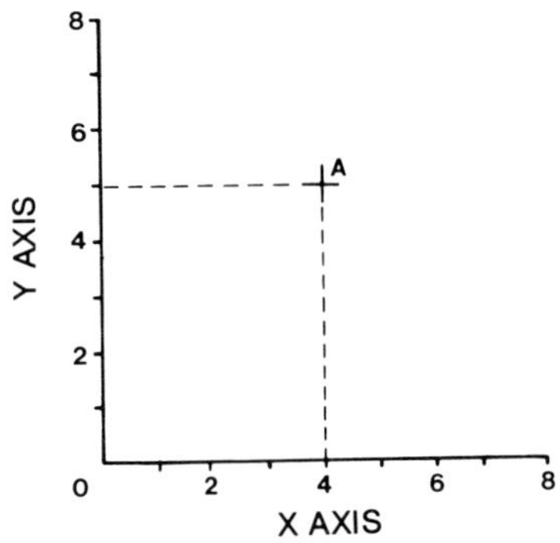

Fig. 3.11. Pair of reference axes

Arrange data in tabular form. Before a graph is plotted, it is often desirable to arrange the data in tabular form. All data to be plotted on the X axis should be arranged in one column, and all data for the Y axis in a parallel column with corresponding items opposite each other (fig. 3.12).

Choose scale and plot points. The scale chosen for a graph depends on the amount of space available and on the type of data to be shown. Scales are normally chosen so that the maximum value along the Y axis uses about 80 to 90 percent of the available space, and the X axis completely fills the available space. A standard graph paper is often used as a base and thus determines the size and scale of the graph. Using the data arranged in tabular form and the scale, the points can be plotted (fig. 3.13).

The next step is to decide what type of graph will be used to show the plotted information.

XYZ DRILLING COMPANY
RIG ACTIVITY

(X AXIS)	(Y AXIS)
Month	No. Rigs
January	13
February	11
March	8
April	12
May	14
June	19

Fig. 3.12. Information arranged in tabular form

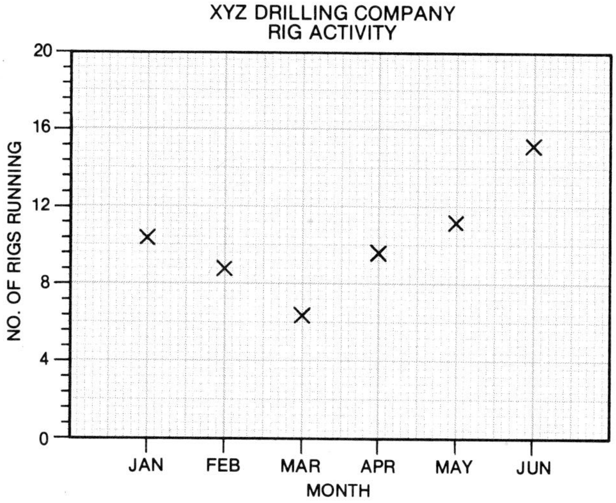

Fig. 3.13. Points plotted for a graph

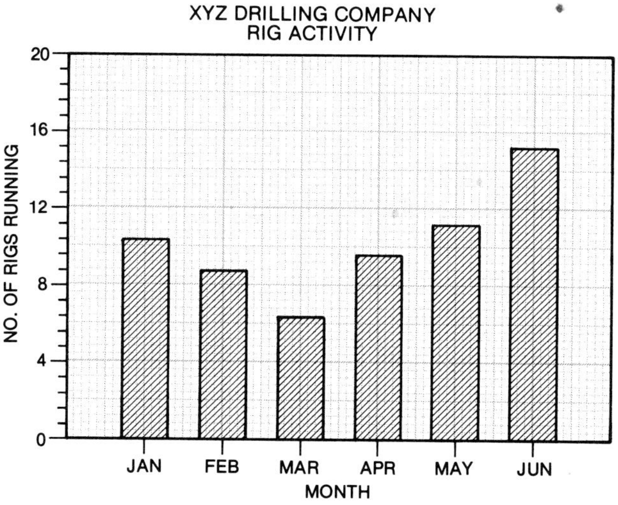

Fig. 3.14. Bar graph showing tabular data given in figure 3.12

TYPES OF GRAPHS

The three types of graphs used most often are the bar graph, the line graph, and the circle graph. Line graphs may have broken lines or curved lines, depending on the data used. Each type of graph offers information at a glance, such as a trend, a comparison, or a curve.

Bar graphs. The bar graph is the preferred way of presenting data representing a series of observations made at periodic intervals. For example, the bar graph in figure 3.14 depicts the tabular information given in figure 3.12 using the points plotted in figure 3.13. The graph shows the relationship between the number of rigs running in January and the number of rigs running in February, and so on. These are relatively unrelated facts; that is, one is not dependent on the other.

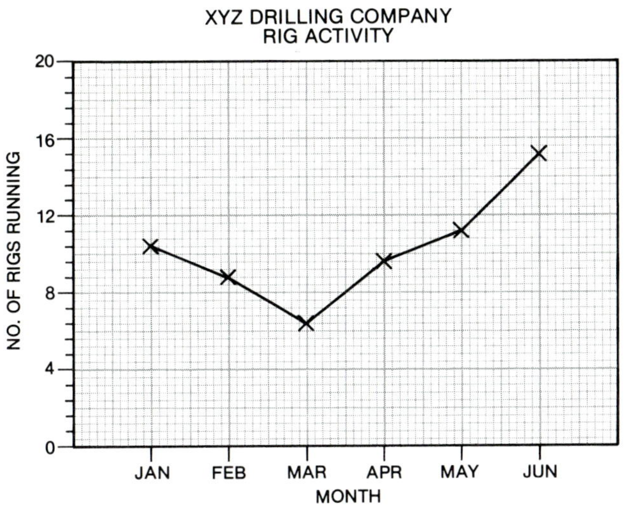

Fig. 3.15. Line graph showing tabular data given in figure 3.12

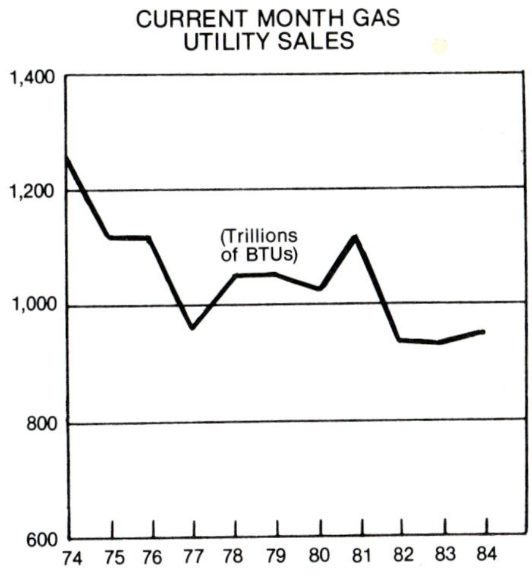

Fig. 3.16. Broken-line graph

Line graphs. The line graph can be used to depict unrelated facts (fig. 3.15); however, it is more commonly used to plot two related variables, such as temperature and time or pressure and time. If the information represents irregular variations, it is plotted by a *broken line* joining the known points (fig. 3.16). If the information represents variables depending directly on each other, a curved line may be used to join the plotted data (fig. 3.17).

Curved-line graphs are also used to show relationships in general—without specific statistics. For example, a plot of vapor pressure for a pure substance would look like that in figure 3.18.

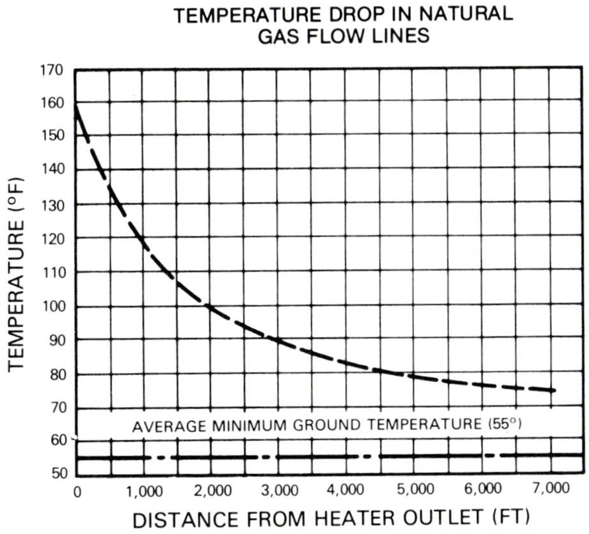

Fig. 3.17. Curved-line graph

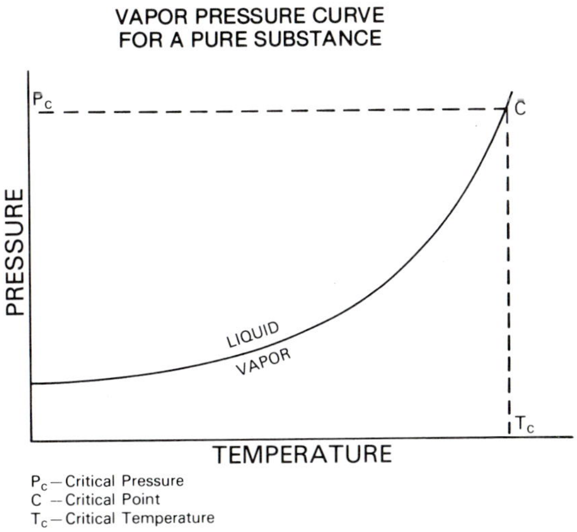

Fig. 3.18. Curved graph showing a general relationship

Circle graphs. A circle graph, commonly called a pie chart, is most often used to compare the various parts of a whole to each other and to the whole. With 360° equaling the whole, the circle is divided so that each 1% of the whole is represented by 3.6°. To construct a circle graph, it is necessary to find the size of the angle for each part. This is accomplished by using the formula

$$\frac{\text{part}}{\text{whole}} \times 360° = \text{angle.}$$

When each angle is determined, it is plotted with a protractor and lines are drawn inside a circle.

Example Problem: A well produces 180 barrels a day. Of these, 90 barrels are oil, 45 barrels are free water, 36 barrels are emulsion, and 9 barrels are sediment. Make a circle graph, or pie chart, to illustrate this distribution.

Solution: Calculate the angles of each "slice of pie" using the formula:

part/whole × 360° = angle.

$90 \div 180 \times 360° = 180°$ (oil)
$45 \div 180 \times 360° = 90°$ (free water)
$36 \div 180 \times 360° = 72°$ (emulsion)
$9 \div 180 \times 360° = 18°$ (sediment)

Then plot the graph as shown in figure 3.19.

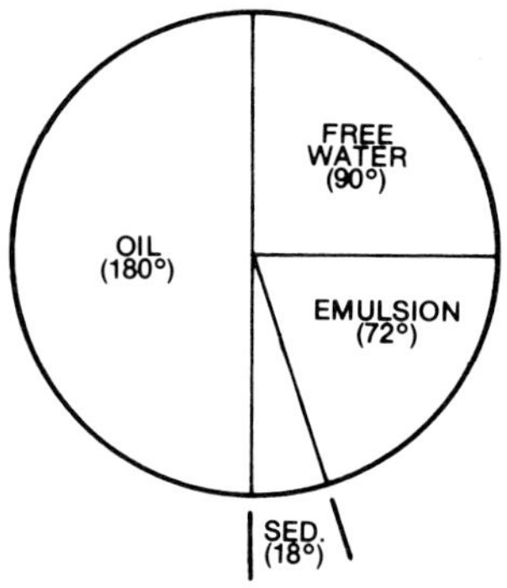

Fig. 3.19. Circle graph, or pie chart

PRACTICE PROBLEMS

1. A worker earned $286.50 one week, of which $60 was spent on rent, $70.80 on food, $7 on laundry, $32.50 on gasoline, $63.40 on incidentals, and the rest was saved. Plot a circle graph showing how his weekly salary was distributed. (Round off degrees to whole numbers before plotting.)

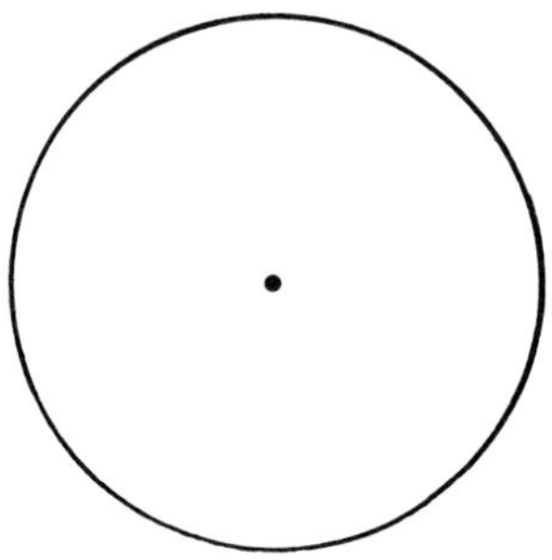

2. Using the graph below as a reference, give *approximate* answers to the following questions.

   a. In what month and year did the excess capacity cease, creating a supply shortage? ________________

   b. In 1981, did the demand for OPEC oil increase or decrease? ________________

   c. How much was the excess capacity of OPEC oil production at the beginning of 1976? ________________

   d. In 1984, what was the average daily OPEC capacity to produce oil? ________________

   e. In what year did OPEC production capacity begin to gradually level out? ________________

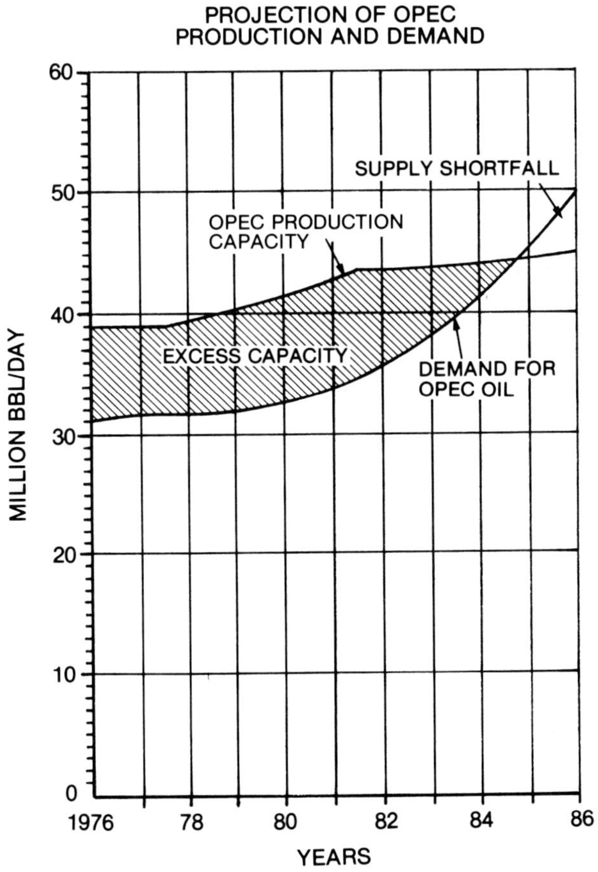

3. The pressure of a certain volume of gas at 50°F is 1,000 psig. At 60°F the pressure increases to 1,020 psig; at 70°F to 1,040 psig; at 80°F to 1,060 psig; and at 100°F to 1,100 psig. Plot a smooth curve from this data on a line graph.

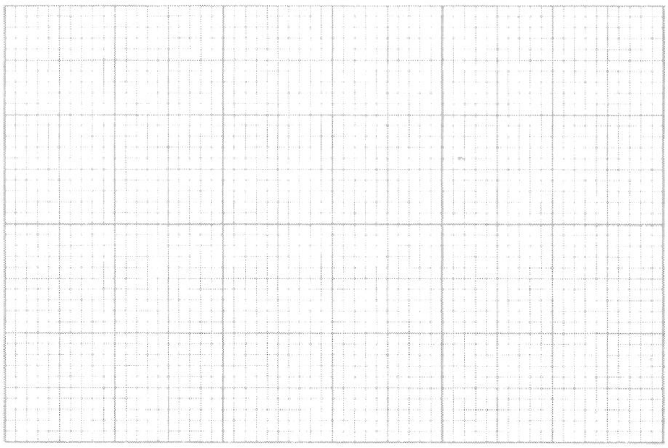

4. The total miles of gas pipe installed by type between 1975 and 1980 will be put into a table. Statistics for plastic pipe installed are 1975—15,985; 1976—11,640; 1977—15,991; 1978—14,576; 1979—18,826; and 1980—18,912. Similar statistics for steel pipe are 1975—12,663; 1976—9,718; 1977—6,157; 1978—7,212; 1979—8,145; and 1980—8,334. Construct a table giving this information.

5. Total world crude production varied from year to year between 1976 and 1980, but offshore crude production showed a steady growth. Using the statistics given in the table at right construct a bar graph that shows total production and offshore production. (Note: Shade a part of each bar to show the offshore production.)

AVERAGE DAILY WORLD CRUDE PRODUCTION (in millions of bbl)

Year	Offshore	Total
1976	8.5	58.5
1977	10.0	56.5
1978	11.0	60.0
1979	12.0	62.5
1980	13.5	59.0

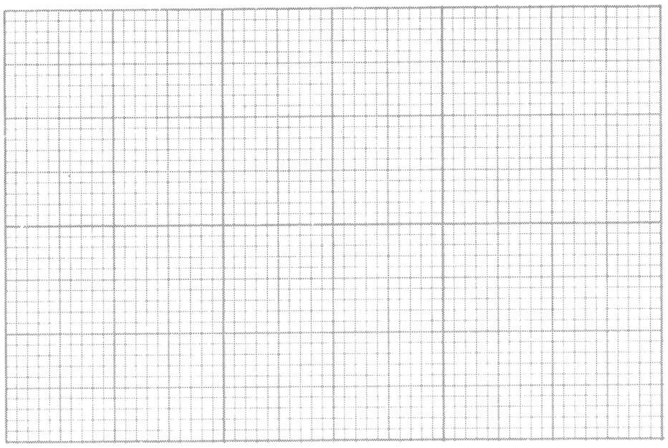

SELF-TEST
3. Number Relations

Multiply each question/problem answered correctly by five to arrive at your percentage of competency.

1. What is the ratio of 25¢ to $1.00? ____________________

2. If ten men can dig 125 feet of ditch in one day, how many feet can twenty-five men dig?

   ____

3. One well produces 75 barrels per day, while a nearby well makes 225 barrels per day. What is the ratio of the production of the first well to the production of the nearby well?

   ____

4. A pickup runs 42 miles on 3 gallons of gas. How many gallons would be needed for a 200-mile trip?

   ____

5. If 16 hammers cost $256, how much will 36 hammers cost?

   ____

6. A man can finish 9 bolts in 12 minutes. How many hours will it take him to finish 165 bolts?

   ____

7. A 6-inch drive pulley makes 800 revolutions per minute. What would be the speed of a 10-inch pulley that is driven by the 6-inch pulley?

   ____

8. An 18-inch pulley on an engine is driving an 11-inch band wheel. The speed for the 18-inch pulley is 180 rpm. What is the speed of the band wheel?

   ____

9. A pinion gear with 16 teeth, running at 1,200 rpm, drives a gear that has 96 teeth. How fast does the gear run?

   ____

10. The U.S. wholesale price of kerosine over a 4-month period was \$42.20, \$42.97, \$39.97, and \$37.71 per barrel. What was the average price per barrel for this period?

____

11. A derrick and a 10-foot pole nearby both cast a shadow as shown below. The derrick shadow is 64 feet, while the pole shadow is 5 feet. How high is the derrick?

____

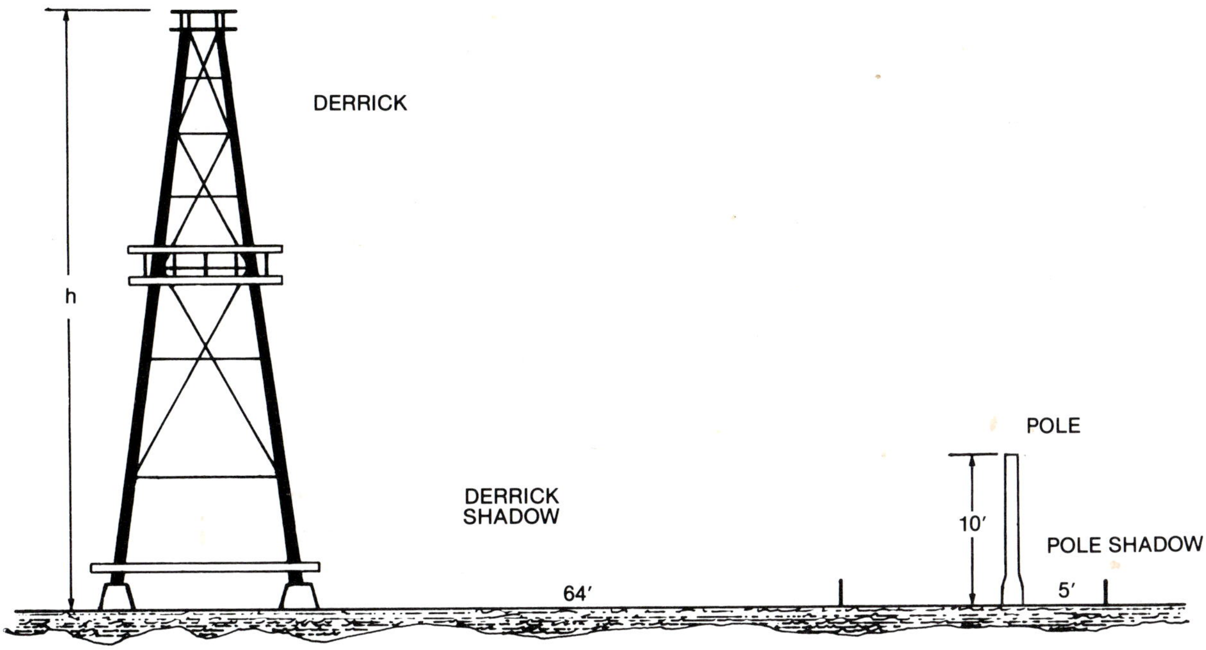

12. Maximum daily temperatures at a plant during one month were (in °F) 46, 47, 51, 50, 45, 48, 61, 59, 55, 54, 47, 52, 55, 46, 39, 43, 52, 56, 49, 51, 53, 56, 63, 51, 54, 59, 58, 49, and 52. Find the mean, median, and mode for this set of temperature data.

Mean ____________________

Median ____________________

Mode ____________________

13. A pipeline buys 1 million cubic feet (MMcf) of natural gas at \$2.67 per thousand cubic feet (Mcf), 48 MMcf at \$2.71 per Mcf, and 36 MMcf at \$2.56 per Mcf. What is the average cost per Mcf?

____

14. The Greens used a total of 7,284 kilowatt-hours (kWh) of electricity for the months designated. The cost per kWh for each month was as shown. Calculate the average price paid per kWh by the Greens during the six-month period.

Month	No. kWh	Cents/kWh
July	1,360	7.50
Aug.	1,424	7.52
Sept.	1,292	7.55
Oct.	1,150	7.50
Nov.	960	7.25
Dec.	1,098	6.97

15. Five quarts of treating solution are added to every 100 barrels of production from a well. How many quarts of treating solution are required for 780 barrels of production?

16. Plot a line graph showing the fluctuation of active rotary drilling rigs during 1982, using the following information:

Month	No. Rigs
Jan.	4,436
Feb.	4,160
Mar.	3,816
Apr.	3,460
May	3,128
June	2,980
July	2,746
Aug.	2,620
Sept.	2,483
Oct.	2,402
Nov.	2,500
Dec.	2,696

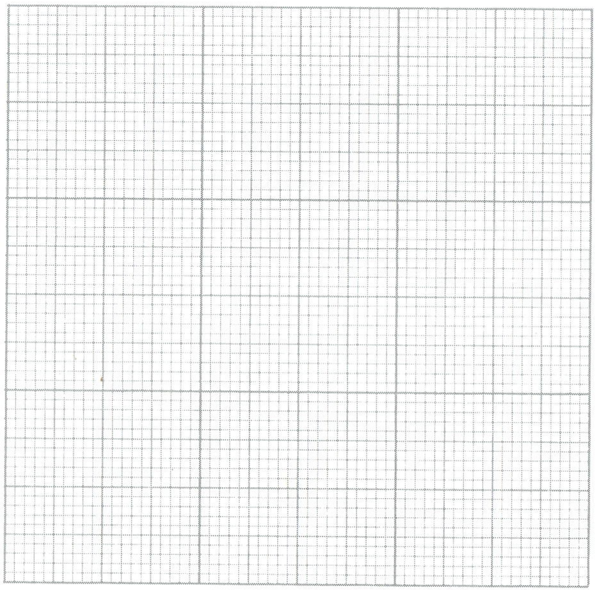

17. Construct a table showing the U.S. wholesale product prices. The prices for gasoline, kerosine, fuel oil, and residual fuel for the last three months in 1982 were as follows: gasoline—$39.68, $38.87, and $36.88 per barrel; kerosine—$42.20, $42.97, and $36.88 per barrel; fuel oil—$40.06, $40.79, and $37.84 per barrel; and residual fuel—$24.78, $26.06, and $24.57 per barrel.

18. Between 1973 and 1982, there was a decline in daily consumption of residual oil in the United States, Japan, and Western Europe, while consumption increased in the rest of the free world. Show this comparison in a bar graph, using the following figures:

	Million Barrels/Day	
	1973	1982
U.S.	2.8	1.7
Japan	2.6	1.4
Western Europe	4.5	2.9
Rest of Free World	3.2	3.6
Total Free World	13.1	9.6

19. Steam-cracking naphtha yielded the following petrochemicals in the percentages shown: ethylene, 23%; propylene, 16%; butene, 9%; butadiene, 5%; fuel gas, 15%; steam-cracked naphtha, 26%; and fuel oil, 6%. Construct a circle graph showing these percentages. Be sure to label each section of the graph. (Round off degrees to whole numbers before plotting.)

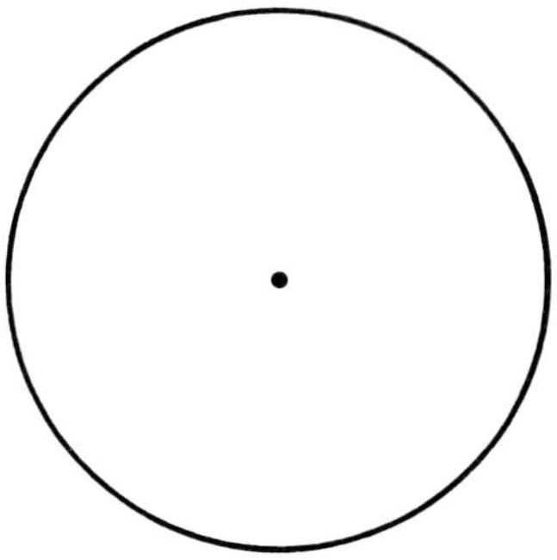

20. The amount of coal produced in the United States between 1970 and 1980 was as follows:

Year	Million Short Tons
1970	612.7
1971	560.9
1972	602.5
1973	598.6
1974	610.0
1975	654.6
1976	684.9
1977	697.2
1978	670.2
1979	781.1
1980	829.7

Present this information graphically.

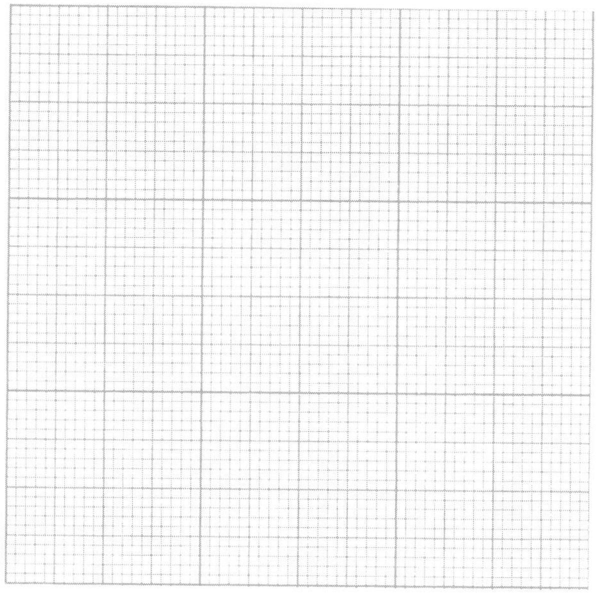

4
SOME PHYSICAL QUANTITIES AND THEIR MEASUREMENT

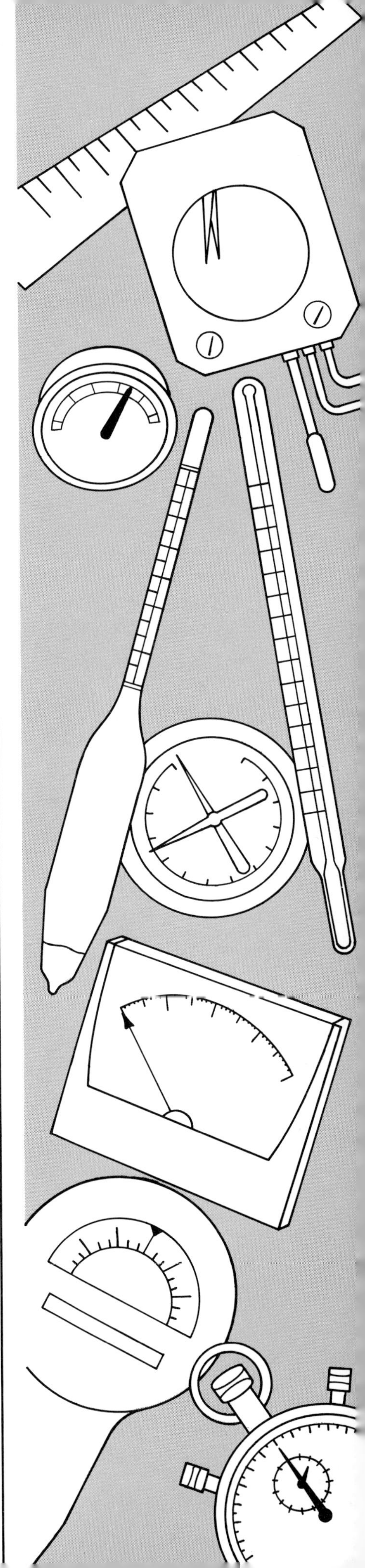

Objectives

Upon completion of chapter four, the student will be able to—

1. Describe the difference between fundamental quantities and derived quantities.
2. Convert measurements from U.S. conventional units to metric units and from metric units to conventional units.
3. Convert units of measurement to equivalent units in order to solve problems.
4. Use and read proper symbols and abbreviations for all common measurement units.
5. Measure distance using a ruler or scale.
6. Solve problems concerning length, area, and volume measurements.
7. Convert temperatures from one scale of measurement to another.
8. Work in all increments of time measurement.
9. Solve problems involving measurements of weight, mass, force, work, power, pressure, density, and specific gravity.
10. Solve problems dealing with the measurement units of electricity including volts, amperes, and ohms.
11. Read simple electrical circuits and solve problems involving voltage, current, and resistance.
12. Solve for kilowatt-hours and other electrical power measurements.

Table 4.1
Conventional and Metric (SI) Units of Measurement
for Some Physical Quantities

Quanitity	Conventional Unit	Conventional Symbol	Metric (SI) Unit	Metric (SI) Symbol
Fundamental Quantities				
Length	foot	ft	metre	m
Mass	pound	lb	kilogram	kg
Time	second	sec	second	s
Temperature	degree Fahrenheit degree Rankine	°F °R	degree Celsius kelvin	°C K
Electrical current	ampere	A	ampere	A
Amount of substance	mole	mol	mole	mol
Luminous intensity	candela	cd	candela	cd
Derived Quantities				
Acceleration	feet per second per second	ft/sec^2	metre per second per second	m/s^2
Area	square feet	ft^2	square metre	m^2
Density	pound per cubic foot	lb/ft^3	kilogram per cubic metre	kg/m^3
Electrical potential	volt	V	volt	V
Electrical resistance	ohm	Ω	ohm	Ω
Electrical capacitance	farad	F	farad	F
Energy	kilowatt-hour	kWh	kilowatt-hour	kWh
Force	pound-force	lb_f	newton	N
Frequency	hertz	Hz	hertz	Hz
Power	foot-pound per second horsepower watt	ft-lb/sec hp W	joule per second watt watt	J/s W W
Pressure	pound-force per square inch	psi	pascal	Pa
Velocity	foot per second	ft/sec	metre per second	m/s
Volume	cubic foot gallon barrel	ft^3 gal bbl	cubic metre	m^3
Work	foot-pound	ft-lb	joule newton-metre	J N·m

Quantity has many definitions. As used here, a *physical* quantity is something that has dimensions and can be measured, such as length, mass (weight), and time. These three physical quantities are among the *fundamental* quantities, those that are not normally divisible into other entities. Additional fundamental quantities are electric current, luminous intensity, temperature, and amount of substance. The seven fundamental quantities are the ones from which all nonfundamental quantities are derived. For example, velocity is a nonfundamental physical quantity since it is composed of distance (length) and time—so many miles per hour, feet per second, and so forth.

The fundamental quantities are also rightfully called *dimensions.* What are the dimensions of this room? That's a good question. You might say, "It's 12 feet wide by 14 feet long by 8 feet high." It is also proper to regard time and mass as dimensions.

A *unit* is a standard measure of a quantity. In the conventional system commonly used in the United States, several units are usually used for each quantity. Miles, yards, feet, inches, and many other units are used for measuring length. For measuring mass (or weight), there are tons, pounds, drams, ounces, and other units. Time is measured in years, months, weeks days, hours, minutes, and seconds. Base units for fundamental quantities, as well as derived quantities, are shown in table 4.1.

In the international system of units (the new metric system, or SI), there is only one unit for each fundamental quantity: *metre* for length, *kilogram* for mass, *second* for time, and so on. Measurements that are much larger or much smaller than these base units are expressed in multiples or submultiples of the fundamental unit, such as kilometres (1,000 metres) and millimetres (0.001 metre).

The metric system enjoys a distinct advantage by using powers of ten and prefix names and symbols. In the conventional system of weights and measures, numerous units may be used to express the value of a fundamental quantity. Typically, these units are clumsily related to other units used to measure the same physical quantity. Inches are divided into quarters, eighths, and sixteenths; a foot has 12 inches; a yard has 3 feet, or 36 inches.

As noted, the metric system has a single unit to represent each fundamental quantity. Measurements much larger or much smaller than the base unit are expressed by multiplying that unit by 10^n, where n is an exponent that can be a positive (+) or negative (−) whole number. A kilometre is 1,000 metres and can be expressed as 10^3 metres. A millimetre is 1/1,000 (0.001) metre, or 10^{-3} metre. A megametre is 10^6 or 1 million (1,000,000) metres. The new metric (SI) encourages the use of ± values of n equal to ±3, ±6, ±9, ±12, ±15, and ±18. These increments allow accurate expression of magnitude and, at the same time, reduce the number of prefixes and symbols required. Table 4.2 lists the names and symbols of the prefixes and their numerical values. The use of deca, deci, hecto, and centi, their symbols, and their numerical values is discouraged by SI but is apt to continue for some time to come.

Table 4.2
Metric Prefixes

Numerical Value	Name	Symbol
*10^1	deca	da
*10^2	hecto	h
10^3	kilo	k
10^6	mega	M
10^9	giga	G
10^{12}	tera	T
10^{15}	peta	P
10^{18}	exa	E
*10^{-1}	deci	d
*10^{-2}	centi	c
10^{-3}	milli	m
10^{-6}	micro	μ
10^{-9}	nano	n
10^{-12}	pico	p
10^{-15}	femto	f
10^{-18}	atto	a

*Not encouraged for use in SI.

Length, Area, and Volume

The most commonly used measurement is length. It is one of the fundamental quantities and is the base unit for area and volume. Area is the product of two lengths, and volume is the product of three lengths.

LENGTH

Length has numerous units by which it can be measured. The more common ones in the conventional system are the *foot* and the *inch*. For measuring small values, the inch is expressed in *mils,* short for thousandths of an inch. In close-tolerance work—for example, measuring the clearance between a bearing cap and a shaft—values of ten-thousandths of an inch are used. The foot is used to measure intermediate lengths or distances; however, a generous overlapping is found in some areas of length measurement. For example, a 3-foot or 3½-foot diameter pipeline is commonly referred to as a 36- or 42-inch pipeline.

Other common units of length in the conventional system are *yards* and *miles.* Yards are famous units on football fields and for measuring amounts of material, such as fabric or carpet. Miles are commonly used to measure distances of travel or distances between points on the earth's surface. *Light years* are used to express the enormous distances that exist in interstellar space. A light year represents the distance light will travel in a year. Since light travels at 186,000 miles per second, a light year is about 5.87 trillion miles.

In the metric system, a simple relation exists among metres, centimetres, and millimetres. Each is easily converted to the others by using multiples or submultiples of 10 as a multiplier. For example, 1 metre equals 10^2 (or 100) centimetres, and 10^3 (or 1,000) millimetres. Conversely, a millimetre equals 10^{-3} (or $1/1{,}000$) metre, and a centimetre equals 10^{-2} (or $1/100$) metre.

Example Problem: Express 0.360 metre in centimetres and millimetres.

Solution: A metre has 10^2 centimetres or 10^3 millimetres. Multiplying by 10^2 requires moving the decimal point two places to the right to obtain centimetres:

$0.360 \times 10^2 = 36.0$ centimetres.

Multiplying by 10^3 means moving the decimal point three places to the right:

$0.360 \times 10^3 = 360$ millimetres.

A disadvantage of conventional measurement is that frequently more than one unit is used to express a measurement. For example, a pipe is

18 feet, 7 inches long. Two units, foot and inch, are used to state the length. That same piece of pipe in metric would be 5 metres, 66 centimetres long, which is simply stated and written as 5.66 metres. This same disadvantage is encountered in weight measurement where typically pounds and ounces are used to express weight.

Example Problem: Express 57 centimetres as a decimal fraction of a metre.

Solution: Since each metre contains 100 centimetres, 57 centimetres is 57/100, or 0.57 metre.

Symbols and abbreviations are useful when dealing with length measurement. A length of 10 feet, 2 inches can be written as 10′2″, or 10 ft, 2 in. Yard and mile are abbreviated as yd and mi, respectively. Metres, kilometres, centimetres, and millimetres are symbolized by m, km, cm, mm, and so on. Metric symbols are never followed by a period. A period after yd, ft, and mi is optional. The abbreviation for inch, *in.*, should have a period to avoid confusion with the word *in.*

Measuring devices for length include rulers, tape measures, surveying chains, and calipers, among others. A surveying chain is a long tape measure, being 100 feet or more in length. A caliper is a tool capable of great accuracy and is generally used to measure rather short distances. A micrometer caliper (fig. 4.1) measures in mils (thousandths of an inch), or it may be calibrated in micrometres (in millionths of a metre).

Small parts of an inch are usually expressed in fractional form: 1/2, 1/4, 1/8, 1/16, 1/32, and 1/64. Sizes smaller than 1/64 are typically measured in mils (0.001 in.) or, for even smaller values, in ten-thousandths (0.0001 in.). Many rulers and tapes are divided into inches and fractional parts as small as 1/16 of an inch (fig. 4.2).

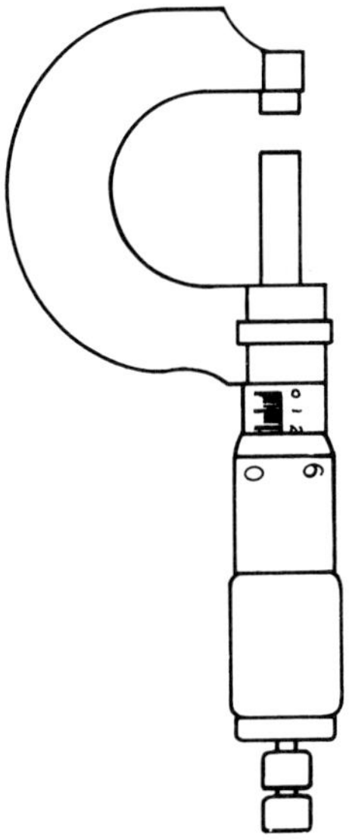

Fig. 4.1. Micrometer caliper

Example Problem: Using figure 4.2, give the distance between *A* and *E* and between *G* and *H*.

Solution:

A to *E* = 2 3/16″.
G to *H* = 7/16″.

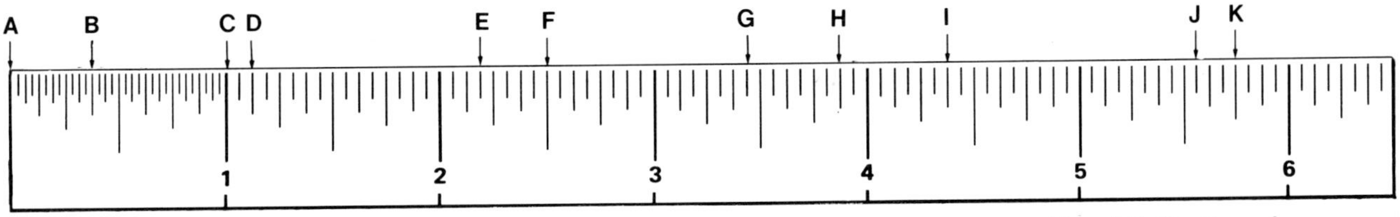

Fig. 4.2. Section of a conventional ruler, divided into sixteenths (actual size)

A ruler graduated in centimetres has each centimetre divided into 10 millimetres. Figure 4.3 shows the true sizes of these small metric units.

Example Problem: Using figure 4.3, give the distance between *A* and *B*.

Solution: Each gradation is 1 millimetre or 1⁄10 of a centimetre, or 0.1 cm. Between *A* and *B*, there are 2 cm and 5 mm. Written as a decimal,

A to *B* = 2.5 cm.

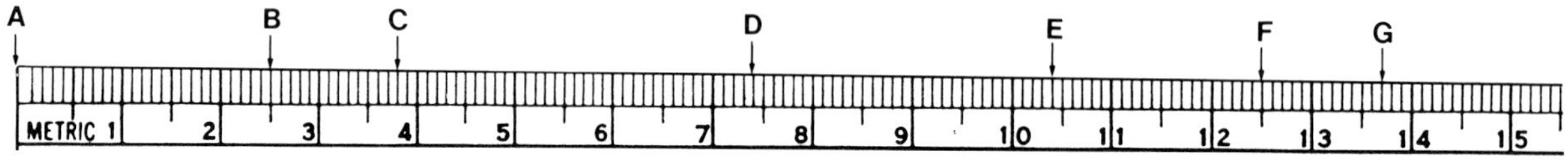

Fig. 4.3. Section of a metric ruler, divided into tenths (actual size)

In addition to converting from metres to centimetres or another metric length measurement, it is often necessary to convert from metric to conventional or vice versa. To convert from conventional to metric units, the metric equivalent is used as a multiplication factor. For example, the metric equivalent of 1 mile is 1.609 kilometres (table 4.3). Thus, to convert 4 miles to kilometres, multiply:

$$4 \times 1.609 = 6.436 \text{ km.}$$

Example: Convert 6 inches to millimetres; then change millimetres to centimetres.

Solution: From table 4.3, find the millimetre equivalent to 1 inch: 25.4. Then multiply:

$6 \times 25.4 = 152.4$ mm.
$152.4 \div 10 = 15.24$ cm.

To convert from metric to conventional, the same principle applies. Find the conventional unit equivalent and multiply this factor by the metric measurement.

Example Problem: A famous gun of World War II was a 155-mm weapon. What was the diameter of this gun in inches?

Solution: Since 1 millimetre equals 0.0394 inch, multiply:

$155 \times 0.0394 = 6.107$.

The diameter of the gun was 6.107 inches.

Table 4.3
Units of Length Measurement

Conventional Unit	Conventional Equivalents	Metric Equivalents
inch (in.)	1,000 mils 0.08333 ft	2.54 cm 25.4 mm
foot (ft)	12 in.	0.3048 m
yard (yd)	3 ft 36 in.	0.9144 m
rod (rd)	5.50 yd 16.5 ft	5.029 m
statute mile (mi)	1,760 yd 5,280 ft	1,609 m 1.609 km
Metric Unit	**Metric Equivalents**	**Conventional Equivalents**
metre (m)	100 cm 1,000 mm	39.37 in. 3.28 ft
kilometre (km)	1,000 m	3,280 ft 1,093 yd 0.62 mi
centimetre (cm)	10 mm 0.01 m	0.394 in.
millimetre (mm)	0.001 m 0.1 cm	0.0394 in.

AREA AND VOLUME

Area and volume are closely related to length. As stated earlier, area is merely the product of two lengths, and volume is the product of three lengths. Sometimes, however, the results of these multiplications are further multiplied by a constant factor, as will be seen.

Examples of area and volume are shown in their simplest forms as a square and a cube (fig. 4.4). Side lengths (a) for both the square and the cube are equal. The area of the square is simply

$$a \times a = a^2,$$

and the volume of the cube is

$$a \times a \times a = a^3.$$

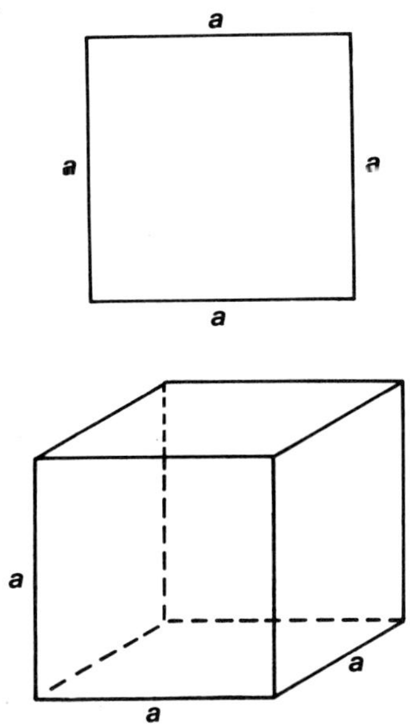

Fig. 4.4 Equal-sided square and cube

If a is equal to 3 feet, the square will have an area of

$$3 \times 3 = 9 \text{ ft}^2,$$

and the cube will have a volume of

$$3 \times 3 \times 3 = 27 \text{ ft}^3.$$

Note that the units of length (feet) are treated as mathematical entities and are squared and cubed, just as the number 3.

Area and volume figures do not have to have equal sides (fig. 4.5). A rectangle has two sides equal to a and two sides equal to b. A rectangular solid has three sides with length a, three with length b, and three with length c. Although sides are of different lengths, the areas and volumes are calculated the same:

$$\text{area} = a \text{ ft} \times b \text{ ft} = ab \text{ ft}^2.$$

$$\text{volume} = a \text{ ft} \times b \text{ ft} \times c \text{ ft} = abc \text{ ft}^3.$$

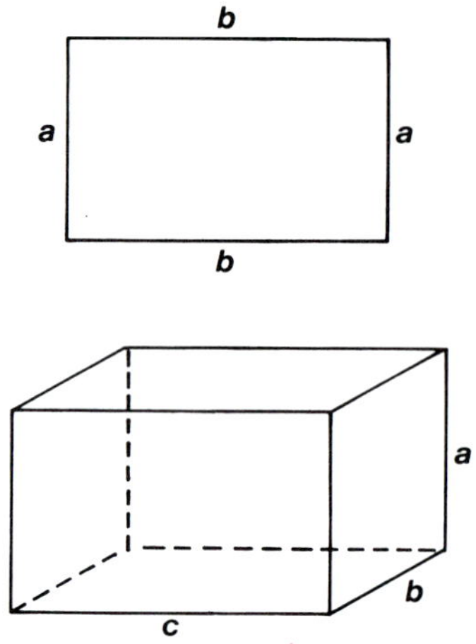

Fig. 4.5. A rectangle and a rectangular solid

A circle has area, and its calculation involves having a product of two lengths. The formula for the area of a circle is πr^2, where r is the radius of the circle, and π (called *pi*) is a constant equal to 3.1416. The product of two lengths is provided by the double use of the radius r; that is,

$$r \times r = r^2.$$

Triangles and, in fact, all plane figures have area, and all are resolved by multiplying two values of length. A simple formula for the area of a triangle is $\frac{1}{2}b \times a$, or $\frac{1}{2}(ab)$ where b is the length of the base and a is the altitude of the triangle (fig. 4.6). Care must be taken to see that measurements of the base and altitude are along lines that are

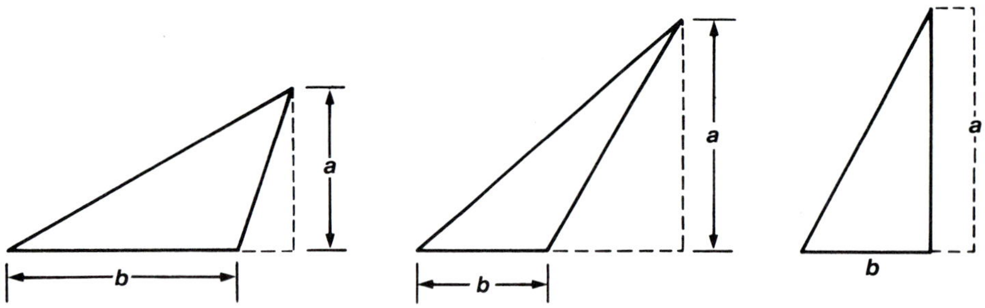

Fig. 4.6. Several forms of triangles

perpendicular to one another as shown by the dotted lines. Some of the logic for establishing the formula for determining the area of a triangle is based on the area of a rectangle. As seen in figure 4.7, a diagonal halving the rectangle divides it into two equal triangles. The area of the rectangle is $a \times b$, so half of that (one of the triangles) would be $\frac{1}{2}(a \times b)$, or $\frac{1}{2}(ab)$.

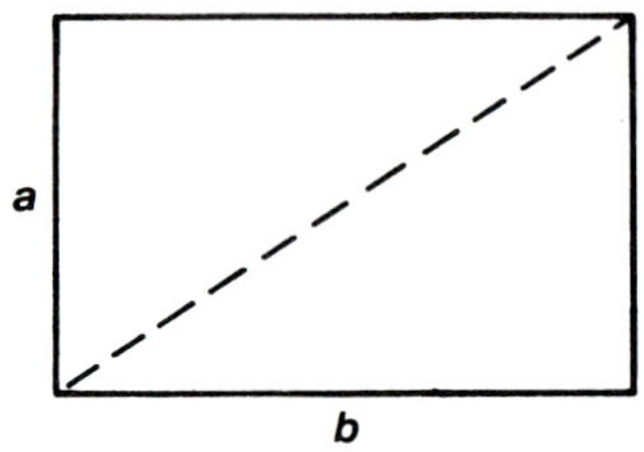

Fig. 4.7. The area of a triangle is equal to half the area of a rectangle.

Areas and volumes can have an unlimited number of shapes and sizes. Both are characterized by having dimensions of length as principal components. In fact, the only other component will be a constant factor, such as ½ (for the area of a triangle), π (for the area of a circle), or $\frac{4}{3}\pi$ (for the volume of a sphere). Formulas for calculating areas and volumes are further explained in chapter six, "Practical Geometry."

Basic conventional units. The square inch, square foot, and square yard are the basic conventional values of *area,* and the cubic inch, cubic foot, and cubic yard are the important basic values of *volume.* In the conventional system of measurements, many units of area and volume have special names, but they are all based on these few units.

Square foot is used commonly to express the floor area in buildings. The fluid ounce, quart, gallon, and barrel are important units of *liquid measure.* The cubic foot is used to measure the production and use of natural gas, as well as for liquids and solids, particularly where the measurement relates to density or specific gravity, as in drilling fluids. The cubic yard is used principally to designate quantities of material such as sand, gravel, and concrete.

The fluid ounce is one of many ambiguous units in the conventional system. The ounce, either troy or avoirdupois, is a measure of weight or mass, while the fluid ounce is a unit of volume. Actually, 1 fluid ounce of water (1.8047 in.^{3}) weighs 1 ounce, avoirdupois, but 1 fluid ounce of alcohol, which is less dense than water, weighs somewhat less.

Metric units. The old metric system had many units for area and volume—square centimetres and square metres for area; and cubic centimetres, litres, and cubic metres for volume. The new version, SI, recognizes only the square metre (m^2) for area and the cubic metre (m^3) for volume. However, the cubic centimetre (formerly cc, now cm^3, or 10^{-6} m^3) and litre are so entrenched in laboratory and commercial use, respectively, that their use will continue for quite a while. The litre is 1,000 cubic centimetres, or 1.0571 U.S. quarts.

Land measurement units. The acre, section, and square mile are important units of land measurement. In the U.S. the standard linear land measures are the mile, rod, and foot. The standard measure for land area is the section or acre. A section is 1 square mile; an acre is 43,560 square feet or 160 square rods. A square acre measures 208.71 feet on each side. Half-sections and quarter-sections are common divisions of land and correspond to 320 acres and 160 acres, respectively. Smaller parcels of land are referred to as a number of acres.

Board measurement units. An interesting and practical measurement peculiar to English-speaking countries is that used for lumber. It is board measurement, and its unit is the *board foot.* A variation of volume measurement, the board foot is described as being 144 cubic inches of wood. A board that is 1 inch thick, 12 inches wide, and 12 inches long contains 1 board foot. Lumber is usually described as having *inches* of thickness, *inches* of width, and *feet* of length, written as 2″ × 4″—12′,

1″ × 12″—16′, and so forth. Fractional values are not considered. A 2″ × 4″ is really closer to 1⅝″ or 1¾″ by 3⅝″ or 3¾″, but board-foot calculations stick with 2″ × 4″. In its simplest form, the equation for board feet is

$$\text{bd ft} = \frac{\text{thickness (in.)} \times \text{width (in.)} \times \text{length (ft)}}{12}$$

Example Problem: Find the board feet in a 2″ × 4″—12′.

Solution:

$$\text{bd ft} = \frac{2'' \times 4'' \times 12'}{12} = 8 \text{ bd ft.}$$

Conversions. Converting from one unit to another is an important factor when solving area and volume problems. Not only is it necessary to change one conventional unit to another, such as square inches to square feet or cubic feet to barrels, but it is often necessary to change from conventional units to metric units and vice versa. Tables 4.4 and 4.5 are designed to help with these conversions. To convert from one conventional unit to another, find the conventional unit of the measurement in the left-hand column of the appropriate table. Then find its equivalent conventional value in the new units desired (middle column) and multiply this figure by the number of original conventional units.

Table 4.4
Units of Area Measurement

Conventional Unit	Conventional Equivalents	Metric Equivalents
Square inch (in.^{2})	0.007 ft^2 0.00077 yd^2	645.16 mm^2 6.451 cm^2
Square foot (ft^2)	144 in.^{2} 0.111 yd^2	0.093 m^2
Square yard (yd^2)	1,296 in.^{2} 9 ft^2	0.836 m^2
Acre	43,560 ft^2 4,840 yd^2	4,046.45 m^2 0.405 hectare
Square rod (rd^2)	30.25 yd^2 0.0006 acre	25.293 m^2
Square mile (mi^2)	640 acres 102,400 rd^2	2.59 km^2
Section	1 mi^2	2.59 km^2
Township	36 mi^2	93.24 km^2

Table 4.5
Units of Volume Measurement

Conventional Unit	Conventional Equivalents	Metric Equivalents
Cubic inch (in.^{3})	0.00058 ft^3 0.0043 gal	16,387 mm^3 16.387 cm^3
Cubic foot (ft^3)	1,728 in.^{3} 7.48 gal	0.028 m^3
Fluid ounce (oz)	1.8047 in.^{3}	29.573 cm^3
Quart	57.75 in.^{3} 32 fluid oz	0.00378 m^3 0.946 litre
U.S. gallon	231 in.^{3} 128 fluid oz 0.1337 ft^3	0.00378 m^3 3.784 litres
Barrel (petroleum)	5.615 ft^3 42 U.S. gal	0.1589 m^3 158.928 litres

Example Problem: How many gallons will a container with a capacity of 532 cubic inches hold?

Solution: Using table 4.5, locate cubic inch in the left-hand column and its equivalent in gallons (0.0043) in the middle column. Multiply this equivalent by 532 cubic inches:

$0.0043 \times 532 = 2.2876$ gal.

The container will hold 2.2876 gallons.

To convert conventional units to metric units, multiply the metric equivalent (right-hand column) by the number of conventional units desired.

Example Problem: How many square metres are there in 8 acres?

Solution: Find acre in the left-hand column of table 4.4, then its metric equivalent (4,046.45 m^2) in the right-hand column. Multiply:

$4{,}046.45 \times 8 = 32{,}371.6$ m^2

There are 32,371.6 square metres in 8 acres.

To convert metric units to conventional units, *divide* the metric units by the metric equivalent.

Example Problem: How many U.S. gallons in 2 cubic metres?

Solution: Table 4.5 shows that 1 gallon has 0.00378 cubic metres, so divide:

$2 \div 0.00378 = 529.1$ gal.

There are 529.1 gallons in 2 cubic metres.

PRACTICE PROBLEMS

Refer to tables 4.3, 4.4, and 4.5 to work these problems.

1. Using figure 4.2, give the dimensions of the following:
   a. *A* to *B* __________
   b. *A* to *K* __________
   c. *C* to *F* __________
   d. *F* to *J* __________
   e. *D* to *I* __________

2. Referring to figure 4.3, give the dimensions of the following in centimetres:
   a. *A* to *D* __________
   b. *A* to *C* __________
   c. *D* to *F* __________
   d. *F* to *G* __________
   e. *B* to *E* __________

3. A football field in the United States is 100 yards between goal lines. How many metres is this distance? __________

4. How many centimetres are in 12 inches? __________

5. Two towns are 32 miles apart. How many kilometres separate the towns? __________

6. A man is 6 feet, 4 inches tall. Express his height in metres. __________

7. Express 240 centimetres in the following units. Show all places.

   a. Metres ______

   b. Millimetres ______

   c. Kilometres ______

   d. Micrometres ______

   e. Megametres ______

8. How many square yards are in 1,000 square metres? ______

9. How long would it take a pump that pumps 50 gallons per minute to fill a 1,000-barrel tank? ______

10. How many cubic inches are contained in 47 gallons? ______

11. How many cubic inches are in a board foot? ______

12. Calculate the board feet in a 1″ × 6″—8′. ______

13. A man is building a patio that will be 8 feet wide by 10 feet long. The concrete forms are 6 inches deep and will be filled to the top. How many cubic yards of concrete will he have to buy? ______

14. How many square miles are in an acre? ______

15. A realtor is dividing a section of land into 2-acre homesites. How many homesites will there be in all? ______

16. How many cubic yards are there in a cubic metre? ______

17. How many litres will it take to make a U.S. barrel of petroleum? ______

18. What cubic-foot capacity is needed to hold 53 gallons? ______

19. A lease containing 530 hectares is how many acres? ______

20. A right-of-way that is 100 yards wide measures 13,492 rods long. How many square miles are contained in the right-of-way? ______

Time and Temperature

MEASURING TIME

Time is measured in a multitude of units, the most common of which are years, months, weeks, days, hours, minutes, and seconds (table 4.6). The year is roughly based on the time required for the earth to orbit the sun. There are 365 + days in a year. (A year has about 6 hours, or one-fourth of a 24-hour day, more than 365 days. The 6 hours are added up and applied every 4 years to form *leap year,* which is 366 days.) Months have various lengths ranging from 28 or 29 days for February to 30 or 31 for the other months.

Each day is 24 hours. A conventional clock is divided into two 12-hour periods—A.M. (midnight to midday) and P.M. (midday to midnight). The military has long used a 24-hour clock with midnight as 0000, noon as 1200, 6 P.M. as 1800, and midnight as 2400 or 0000. The 24-hour clock is gaining popularity rapidly.

Each hour is divided into minutes (60 of them) and seconds (3,600 of them). The second is the fundamental unit of time in SI metric. The standard second is called the *ephemeris* second. At one time the second was defined in terms of the solar year, but today the ephemeris second is defined in terms of the frequency of radiation from the cesium 133 atom. *Ephemeris* is a word that describes a chart, table, or almanac related to astronomy.

Seconds, minutes, hours, and days are the important units for describing the time rate at which something is being accomplished—how many revolutions a motor makes per minute (rpm), how much fuel an aircraft consumes per hour of flight, and how many miles per hour a car travels are examples.

Very short intervals of time are measured in milliseconds (ms or 10^{-3} s), microseconds (μs, or 10^{-6} s), nanoseconds (ns, or 10^{-9} s), and so on.

Table 4.6
Common Units of Time

Unit	Equivalent Values
Century	100 years
Decade	10 years
Year	12 months 365 + days* 8,760 + 6 hours*
Week	7 days 168 hours 10,080 minutes
Day	24 hours 1,440 minutes
Hour	60 minutes 3,600 seconds
Minute	60 seconds
Second	0.0167 minute

*A year has 365 days plus 6 hours. The 6 hours are added up and applied every 4 years to form leap year, which has 366 days.

Example Problem: An electric motor turns at 1,800 revolutions per minute (rpm). How many revolutions does it make each second?

Solution: Set up a proportional equation with x representing the unknown:

$$1{,}800/60 = x/1$$
$$60x = 1{,}800 \times 1 = 1{,}800$$
$$x = 1{,}800 \div 60 = 30.$$

The motor makes 30 revolutions each second.

TEMPERATURE MEASUREMENT

The measurement and control of temperature are important aspects of everyday life as well as industry. Temperature, like pressure, forms a part of everyone's environment. Up to this time, absolute zero values of these two physical quantities have not been achieved, although zero values have been closely approached.

In the conventional measurement system, temperature is expressed in degrees Fahrenheit (°F). Important points on the Fahrenheit scale are the temperature of water at its freezing point (32°F) and its boiling point (212°F). Worldwide, the popular temperature scale is Celsius (°C). The Celsius scale supplants the centigrade scale, although the two are identical. With the introduction of the new metric system, SI, a new name was needed because the word *centigrade* contains an important prefix (centi) and its use introduces an awkward anomaly into the system of units.

The kelvin scale, popularly called the *absolute* temperature scale for metric measurements, has its zero at −273.16°C. The degree sign (°) is not used with kelvin, and the term *kelvin* is not written with an uppercase *K,* although an uppercase K is the symbol for kelvin.

Another absolute temperature scale is the Rankine (R). Based on the Fahrenheit scale, it serves the conventional system of measurements.

The Fahrenheit and Celsius scales are in common use for everyday functions such as weather forecasting, heating and cooling, cooking, and medical care. The absolute scales are required in many applications, particularly those dealing with physics and chemistry. For example, the various laws relating to behavior of gases require the use of absolute scales.

Figure 4.8 shows the relationship of all four temperature scales. These scales are related through fairly simple equations as shown in table 4.7.

Example Problem: Convert 250°F to Celsius.

Solution: Use the formula for converting Celsius to Fahrenheit.

$$°C = 5/9(°F - 32°) = 5/9(250°F - 32°) = (5 \times 218) \div 9 = 121.11$$

or

$$°C = (°F - 32°)/1.8 = (250°F - 32°) = 218 \div 1.8 = 121.11.$$

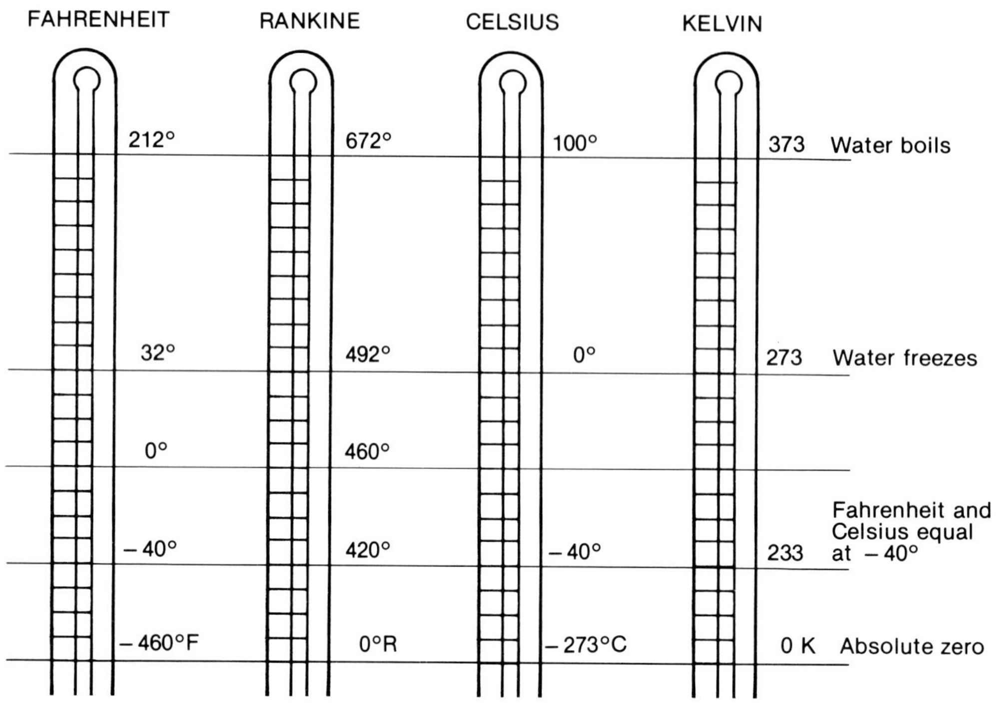

Fig. 4.8. A comparison of the various temperature scales. Note that Celsius and Fahrenheit scales are equal when each registers −40°.

Table 4.7
Temperature Scales

Scale	Equivalents
°F	9/5°C + 32, or 1.8°C + 32 9/5(K − 273) + 32, or 1.8(K − 273) + 32 °R − 460
°C	5/9(°F − 32), or (°F − 32)/1.8 K − 273 5/9(°R − 492), or (°R − 492)/1.8
°R	°F + 460 9/5K, or 1.8K 9/5°C + 492, or 1.8°C + 492
K	°C + 273 5/9°R, or °R/1.8 5/9(°F − 32) + 273, or (°F + 460)/1.8

PRACTICE PROBLEMS

Refer to tables 4.6 and 4.7 to solve these problems.

1. An airliner traveled 3,975 miles from Dallas to Honolulu in 7½ hours.

   a. What was its average speed in miles per hour? __________

   b. What was its speed in miles per minute? __________

2. The capacity of a pump is 1,000 gallons per minute. How long will it take to fill a 500-barrel tank? (42 gal = 1 bbl.) __________

3. For every 90 revolutions of a pumping engine, the pump rods are raised and lowered once. If there are 15 strokes per minute, what is the rpm of the engine? __________

4. Convert the following temperatures to Fahrenheit:

   a. 345 K __________

   b. 20°C __________

   c. 0°C __________

   d. 482°R __________

   e. −78°C __________

5. Convert the following temperatures to Celsius:

   a. −89°F __________

   b. 32°F __________

   c. 277 K __________

   d. 432°R __________

   e. 10 K __________

6. What are the Celsius temperature values for the following kelvin values?

   a. 324 K __________

   b. 203 K __________

   c. 293 K __________

   d. 250 K __________

   e. 49 K __________

7. What are the conventional absolute temperature values for the following Fahrenheit temperatures?

   a. 680°F __________

   b. 3°F __________

   c. 43°F __________

   d. −24°F __________

   e. −113°F __________

8. In batching products through a pipeline, the rate of travel of the fluid through a 50-mile pipeline is 1.8 miles per hour. A new batch is started at 9:00 A.M. on a Tuesday. When will this batch first arrive at the terminal end? __________

9. At a loading dock for oil tankers the oil can be delivered to a tanker at the rate of 10,000 barrels per hour. How many minutes will it take to load a 16,500-barrel tanker? __________

10. Convert 325 degrees Rankine to the metric absolute temperature equivalent. __________

Mass and Related Derived Quantities

Mass, along with length and time, constitutes one of the fundamental quantities on which all physical measurements are based. It is that property of a body that is commonly considered as the amount of material it contains, that causes it to have weight in a gravitational field, and that determines its inertia. Derived quantities closely related to mass are weight and force, density and specific gravity, work and power, and pressure.

WEIGHT AND FORCE

In conventional terms weight and mass are considered to be the same thing, but sometimes weight and force are treated as being the same thing. In oilwell and gas-well drilling operations, a common expression is "weight on bit." If metric units are used in the operation, the weight on bit is likely to be expressed in newtons or decanewtons, which are units of force. If conventional units are used, the weight will be expressed in pounds-force, usually stated as just "pounds."

In the United States, the pound-mass is the unit of mass and is symbolized as lb_m to distinguish it from the pound-force, lb_f. The pound-mass is defined in terms of the metric system unit of mass, the kilogram. The pound-mass is 0.45359237 kilogram. The U.S., or conventional, pound-force is defined to be the force necessary to accelerate a mass of 1 pound at 32.15 feet per second squared (32.15 ft/s^2). The metric unit of force, the newton, is the force needed to accelerate a mass of 1 kilogram at 1 metre per second squared ($kg \cdot m/s^2$). Table 4.8 lists the most-used units of mass, weight, and force in the conventional and the metric systems.

Example Problem: A spring is acting against a heavy door with a force of 30 decanewtons (daN). What is the force in pounds-force (lb_f)?

Solution: A newton is equal to 0.2248 pounds-force (from table 4.8), and a decanewton is 10 newtons, or

$$10 \times 0.2248 = 2.248 \text{ lb}_f.$$

Then 30 daN would be

$$30 \times 2.248 = 67.44 \text{ lb}_f.$$

DENSITY AND SPECIFIC GRAVITY

Density. Density is defined as the weight of a substance per unit volume. A unit volume can be a cubic inch, a cubic foot, a gallon, a barrel, a

Table 4.8
Units of Weight, Mass, and Force Measurement

Conventional Unit	Conventional Equivalents	Metric Equivalents
ounce (oz)	0.0625 lb_m	28.35 g
pound-mass (lb_m)	16 oz	0.4536 kg 453.6 g
ton (tn)	2,000 lb_m	0.9072 t 907.185 kg
pound-force (lb_f)	$lb_m \times 32.15\ ft/s^2$	4.448 N
Metric Unit	**Metric Equivalents**	**Conventional Equivalents**
gram (g)	0.001 kg	0.03527 oz.
kilogram (kg)	1,000 g	2.2046 lb_m
tonne (t)	1,000 kg	1.10231 tn
newton (N)	$kg{\cdot}m/s^2$	0.2248 lb_f

cubic yard, or any of a number of measurement units. The weight, or mass, can be any of the units studied, although they properly should be in the same system of units as the volume—that is, either metric or conventional. In conventional units, pounds and cubic feet are the most commonly paired units, although pounds per gallon is used to measure density in some instances, particularly when dealing with liquids.

In the metric system, grams per cubic centimetre (g/cm^3) is in common use, although in SI only kilograms per cubic metre (kg/m^3) is recognized as a density measure. A cubic centimetre is $(10^{-2}\ m)^3$, or $10^{-6}\ m^3$, while a gram is 10^{-3} kg, so

$$g/cm^3 = 10^{-3}/10^{-6}\ m^3 = 10^3\ kg/m^3 = 1{,}000\ kg/m^3.$$

One gram per cubic centimetre and 1,000 kilograms per cubic metre are equal quantities as far as density is concerned and represent the density of pure water at a temperature of 4° Celsius.

Density as a characteristic of materials is an important consideration because it provides a clue as to how heavy a finished product might be. For example, when density and strength are both taken into account, the proper choice of material for a particular application can be made. Aluminum alloy is favored over steel as a material for aircraft use because it has less density than steel for a given strength. This means a lighter aircraft of adequate strength can be produced. On the other hand, in many instances in aircraft construction, steel is chosen over aluminum alloy because steel has advantages over other metals for certain applications.

The weight of a large body may be calculated if the density of the material and the volume of the body are known. Densities of certain solids, liquids, and gases have been calculated and are given in table 4.9. The density of each is given in pounds per cubic foot (lb/ft^3), kilograms per cubic metre (kg/m^3), and grams per cubic centimetre (g/cm^3).

Example Problem: The volume of a gasoline storage tank is 5,360 cubic feet. What is the weight of the gasoline it will contain when full?

Solution: From table 4.9, find the density of gasoline in pounds per cubic foot. Then set up the proportion, 46.8 pounds is to 1 cubic foot as x pounds is to 5,360 cubic feet:

$$46.8/1 = x/5{,}360$$
$$x = 46.8 \times 5{,}360 = 250{,}848.$$

The gasoline will weigh 250,848 pounds.

Table 4.9
Densities of Some Solids, Liquids, and Gases

Material	lb/ft^3	kg/m^3	g/cm^3 (sp gr)
Solids			
Gold	1,206.2	19,300	19.3
Mercury	846.0	13,500	13.5
Lead	712.5	11,400	11.4
Iron	485.0*	7,700*	7.7*
Aluminum	165.6	2,600	2.6
Wood	50.0*	800*	0.8*
Ice	56.9	900	0.9
Liquids			
Sulfuric acid	125.0	2,000	2.0
Seawater	64.3	1,030	1.03
Fresh (pure) water	62.5	1,000	1.00
Kerosine	50.0	800	0.80
Gasoline	46.8	750	0.75
Gases			
Air	0.075	1.20	0.0012
Oxygen	0.084	1.34	0.00134
Nitrogen	0.0737	1.18	0.0018
Carbon monoxide	0.0734	1.17	0.0017
Hydrogen	0.0053	0.085	0.000085

*Wood and iron have variable densities. The values shown are approximate figures.

Specific gravity. Specific gravity (sp gr) is the ratio between the weight of a given volume of a substance and the weight of an equal volume of pure water:

$$\text{sp gr of sub.} = \frac{\text{wt/vol of substance}}{\text{wt/vol of pure water}}$$

or

$$\text{sp gr} = \frac{\text{density of substance}}{\text{density of pure water}}$$

Like density, the weight per volume can be expressed in lb/ft^3, kg/m^3, or g/cm^3. When the density is expressed in grams per cubic centimetre, the specific gravity of a substance is the same as its density because the density of pure water is 1 gram per cubic centimetre, and any number divided by 1 equals the same number. Thus, the densities given under the column headed "g/cm^3" in table 4.9 are also the specific gravities for those substances. In the metric system (SI), specific gravity is called *relative density.*

Example Problem: What is the specific gravity of a drilling mud that weighs 78.54 lb/ft^3?

Solution: From table 4.9, the density of pure water is 62.5 lb/ft^3.

$$\text{sp gr} = 78.54 \div 62.5 = 1.257.$$

The same equation may be used to find the weight or density of a substance by transposing it to read:

$$\text{density of substance} = \text{density of pure water} \times \text{sp gr}$$

or

$$\text{wt/vol of substance} = \text{wt/vol of pure water} \times \text{sp gr.}$$

The specific gravities of gases are frequently expressed as a ratio between the weight of a given volume of the gas and the weight of an equal volume of air. Specific gravity of gases expressed in this manner may also be referred to as *vapor density.*

Table 4.10 shows specific gravities based on air for several gases.

Table 4.10
Specific Gravity
Based on Air

Gas	Sp gr
Air	1.00
Oxygen	1.120
Nitrogen	0.0983
Carbon monoxide	0.979
Hydrogen	0.071
Butane	2.004

Baumé scales. Several scales based on specific gravity have been developed for specialized use. Two Baumé scales have been in use for centuries—one for measuring heavy liquids, the other for measuring gravity of liquids lighter than fresh water. The increments in gravity are called degrees Baumé (°Baumé). Two equations express the two scales:

1. For heavy liquids: $°\text{Baumé} = 145 - 145/\text{sp gr}.$
2. For light liquids: $°\text{Baumé} = 140/\text{sp gr} - 130.$

Note that in equation 1 if the liquid has a specific gravity of 1, as it would for water, the degrees Baumé will be 0, but in equation 2 it will be 10. In equation 2 the degrees Baumé increase as the specific gravity decreases. In equation 1, the degrees Baumé increase as the specific gravity increases.

°API gravity scale. During the early 1920s the American Petroleum Institute (API), the Bureau of Mines, and the National Bureau of Standards adopted a special gravity scale for measuring crude petroleum and some of its products. It is designated °API and bears a resemblance to the Baumé scale for light liquids:

$$°\text{API} = (141.5/\text{sp gr}) - 131.5.$$

The advantage of the Baumé and API scales lies in the expansion that is obtained between very small changes in specific gravity. For example, a change from 28°API to 30°API is equivalent to a change from 0.8348 to 0.8251 in specific gravity. The API scale is clearly easier to use.

To solve for specific gravity when the °API gravity is known, the equation can be transposed to read:

$$\text{sp gr} = \frac{141.5}{°\text{API} + 131.5}$$

Example Problem: What is the specific gravity of 43°API gravity oil?

Solution:

$$\text{sp gr} = 141.5 \div (43 + 131.5) = 0.810882 = 0.811.$$

PRESSURE

Pressure is defined as force per unit area. The force can be pounds-force or newtons. The area can be square inches, square centimetres, square metres, and so on. *Pounds per square inch* (psi) is the favorite conventional unit of pressure. The metric (SI) demands *newton per square metre* and calls it a *pascal* (Pa). The pascal is a small value of pressure; one psi equals 6,895 pascals, or 6.895 kilopascals (kPa).

Another measure of pressure that does not agree with the definition of force per unit area is *kilograms per square centimetre.* Kilogram is the unit of mass in SI, and its use as a force unit is not allowed. However, it is a popular measure of pressure and a fairly large unit. A kilogram-force per square centimetre is about 98 kPa, or about 14.2 psi. Another form of pressure measurement is the height of a liquid column, typically a column of mercury or water.

Like temperature, there is an absolute scale of pressure. The air that surrounds the earth exerts a pressure of about 14.7 pounds per square inch at sea level. This is the pressure value *above* absolute zero pressure, and it is designated as *pounds per square inch, absolute,* or psia. In metric the value is about 101.3 kPa, and for mercury columns it is about 760 mm or 29.92 inches.

A typical pressure gauge shows *gauge pressure* (psig), meaning that the atmospheric pressure of 14.7 corresponds to *zero* on the pressure gauge. Readings below atmospheric pressure may be shown on the gauge, and the values will probably be expressed in inches of mercury, vacuum. In converting gauge pressure to absolute pressure, the formula is

$$\text{psia} = \text{psig} + 14.7.$$

Physical laws of gases state that absolute pressure of a gas varies directly with its absolute temperature as

$$P_2/P_1 = T_2/T_1$$

where P and T signify pressure and temperature *before* (P_1, T_1) and *after* (P_2, T_2) a change. This formula can be applied only where the volume of gas does not change.

Example Problem: Suppose the pressure on a 10 ft^3 tank of air was 30 psig at 70°F. After the sun had shone on it all morning, the temperature rose to 110°F. How much did the pressure increase?

Solution: Convert both temperatures to absolute values:

$70° + 460° = 530°.$
$110° + 460° = 570°.$

Convert the pressure to absolute value:

30 psig + 14.7 = 44.7 psia.

Then

$P_2 \div 44.7 = 570 \div 530$
$P_2 = 44.7 \times 570 \div 530 = 48.07 = 48.1$ psia.

The pressure increase is

$48.1 - 44.7 = 3.4$ psi.

Hydrostatic pressure is the force exerted by a body of fluid at rest and increases directly with the density and the depth of the fluid. The hydrostatic pressure of fresh water is 0.433 psi per foot of depth, or 9.792 kPa per metre of depth.

Example Problem: How much hydrostatic pressure is exerted by a 150-foot column of water?

Solution:

$150 \times 0.433 = 65$ psi.

WORK AND POWER

Work is the result of force that produces motion and is expressed as a product of force and distance. In the conventional system of measurements, the commonly used unit of force is the pound, and the common distance unit is the foot, so work is normally expressed in foot-pounds (ft-lb). For example, if a can of fuel weighing 35 pounds is lifted 3 feet to the bed of a truck, the amount of work done is

$$3 \text{ ft} \times 35 \text{ lb} = 105 \text{ ft-lb}.$$

Example Problem: A man pushes a hand truck with a force of 50 pounds a distance of 20 feet. How much work does he perform?

Solution:

$20 \text{ ft} \times 50 \text{ lb} = 1{,}000 \text{ ft-lb}$ of work.

Power is the time rate of doing work. Power is expressed in foot-pounds per unit of time, usually minutes but sometimes seconds or hours. Foot-pounds per minute and foot-pounds per second are common expressions for power. The standard unit for mechanical power is the horsepower (hp), adopted centuries ago. It was established that an average horse could do 33,000 foot-pounds of work per minute, and that became the basis for the standard equation

$$\text{hp} = \text{ft-lb/min} \div 33{,}000.$$

Example Problem: What horsepower does a pump that lifts 5,000,000 foot-pounds of water per hour expend?

Solution: Since this amount is for a period of 1 hour, find the foot-pounds per minute (60 min = 1 hr):

$5{,}000{,}000 \div 60 = 83{,}333.33 \text{ ft-lb/min}.$

Then, using the equation for finding horsepower:

$83{,}333.33 \div 33{,}000 = 2.525 \text{ hp}.$

The unit of work and energy in the metric system is the newton-metre (N·m) and is called the *joule* (J). One foot-pound in the conventional system equals 1.356 joules, and 1 joule equals 0.737 foot-pound. A force of 1 newton acting through a distance of 1 metre equals 1 joule.

The unit of power in the metric system is the joule per second (J/s) and is called the *watt* (W). The larger unit of power, the *kilowatt* (kW), is a world standard for electric power, and the kilowatt-hour is a standard for energy use. A kilowatt-hour is equal to 3.6 megajoules as shown by the equation

$$1 \text{ kW} = 1{,}000 \text{ J/s} \times 3{,}600 \text{ s} = 3{,}600{,}000 \text{ J} = 3.6 \text{ MJ}.$$

One horsepower is equal to 746 watts, or 0.746 kilowatts.

Example Problem: How many megajoules (MJ) are in 75 horsepower-hours?

Solution: First convert horsepower-hours to kilowatt-hours:

75 hp-hr × 0.746 kWh = 55.95 kWh.

Then convert to megajoules:

55.95 kWh × 3.6 MJ = 201.42 MJ.

PRACTICE PROBLEMS

Refer to tables 4.8, 4.9, and 4.10 to solve these problems.

1. How many kilograms are in a conventional ton? __________
2. The weight on bit at a drilling rig is 120,000 lb_f. Express that force in decanewtons (daN) and kilonewtons (kN). __________
3. A truck load capacity is 6 tons. How many feet of 6-inch pipe can it haul if the pipe weighs 19 pounds per foot? __________
4. How many pounds are in 5 kilograms? __________
5. How much more does a cubic foot of seawater weigh than an equal volume of fresh water? __________
6. If the specific gravity of a certain crude is 0.82140, what is its °API gravity? __________
7. A man weighing 160 pounds climbs a 20-foot ladder. How much work has he done in the climb? __________
8. Convert a gauge pressure of 36.9 psi to absolute pressure. __________
9. What is the difference in the weight of a barrel of 40°API gravity oil and one of 50°API gravity oil? (1 bbl = 5.6 ft^3) __________
10. What is the pressure at the bottom of a 1,600-barrel water storage tank when the height of the water is 16 feet? __________

Electricity

Electricity is the flow of electrons through a conducting medium, such as a metal wire. The flow of electricity is much like the flow of fluid in a pipe. For one thing, pressure is needed to force the fluid to flow—that is, there must be a pressure difference between one point in the pipe and another point if the fluid is to flow between the two points. The same applies to electron flow—a pressure difference is needed between two points in a wire or other conducting material.

In electricity, the pressure difference is called *voltage, potential,* or *potential difference.* Two other important quantities in simple electrical circuitry are *current* and *resistance.* Just as voltage is like pressure in fluid flow, current is like the rate of fluid flow, and resistance is like friction in a fluid piping system.

The standard unit of measurement for voltage is the *volt* (V); for current, the ampere (A); and for resistance, the ohm (Ω) (table 4.11). Fortunately, the science of electricity was developed largely after the French scientists had introduced the metric system. Consequently, units in electricity are the same in both the conventional and metric systems.

Table 4.11
Basic Electrical Units of Measure

Quantity	Symbol	Unit Name	Unit Symbol
Voltage	E	volt	V
Current	I	ampere	A
Resistance	R	ohm	Ω

OHM'S LAW

Ohm's law states that the strength or intensity of an unvarying electrical *current* is directly proportional to the electromotive force (*voltage*) and inversely proportional to the *resistance* of the circuit. Simplified, Ohm's law shows the relationship among the three electrical quantities to be

$$I = E/R$$
$$E = I \times R$$
$$R = E/I$$

where

[redacted]
I = current in amperes
E = voltage in volts
R = resistance in ohms.

An easy way to remember Ohm's law is to use the Ohm's law wheel (fig. 4.9). To use the wheel, cover the unknown factor and read the resulting equation. Suppose the unknown factor is I. Cover I and the diagram indicates that I equals E/R. When E is the unknown factor, the equation is $E = IR$, and when R is the unknown, the equation is $R = E/I$.

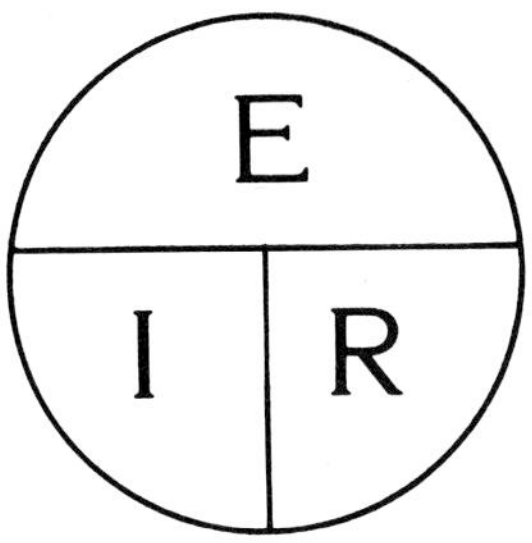

Fig. 4.9. Ohm's law wheel

Ohm's law is useful in determining (1) the amount of current that will flow in a circuit, (2) the amount of resistance needed in a circuit to limit current to a given value, or (3) the voltage needed to cause a given flow of current through a particular value of resistance.

Example Problem: What voltage is needed to force 1.2 amperes through a device that has a resistance of 80 ohms?

Solution: Use the formula for finding voltage E:

$$E = IR.$$

$$I = 1.2 \text{ amperes}; \; R = 80 \text{ ohms}.$$

$$E = 1.2 \times 80 = 96 \text{ volts}.$$

Other relations exist among the three quantities, but $I = E/R$ is the basis for Ohm's law. The quantity *power* is also derived from this formula, as described later.

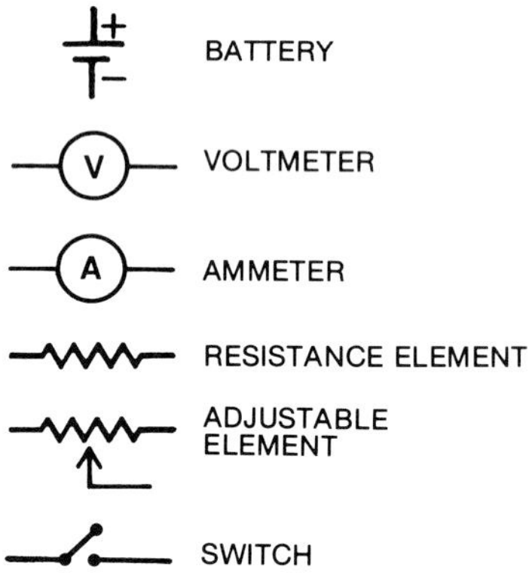

Fig. 4.10. Electrical symbols

CIRCUITS

Kinds of electricity include static, direct current, and alternating current. Both direct-current and alternating-current electricity are of great importance in everyday life, but in this section only direct-current electricity will be considered. It can be obtained from various kinds of cells and batteries, as well as from generators.

In the discussion of circuits, various symbols are used for electrical components (fig. 4.10). These symbols are standard means of representation and are commonly used in the drawing of circuits.

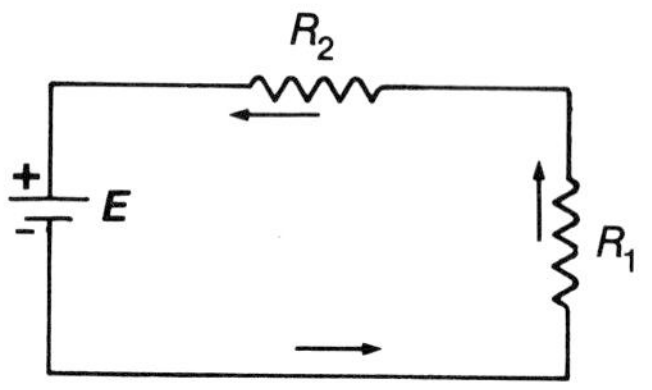

Fig. 4.11. Series circuit

Two popular types of circuits for direct current are designated as series (fig. 4.11) and parallel (fig. 4.12). In the series circuit, two resistors are in series, and the total current flows through each resistor (R). In the parallel circuit, the resistors are connected in parallel, and the total current divides, with part going through one resistor and part going through the other resistor.

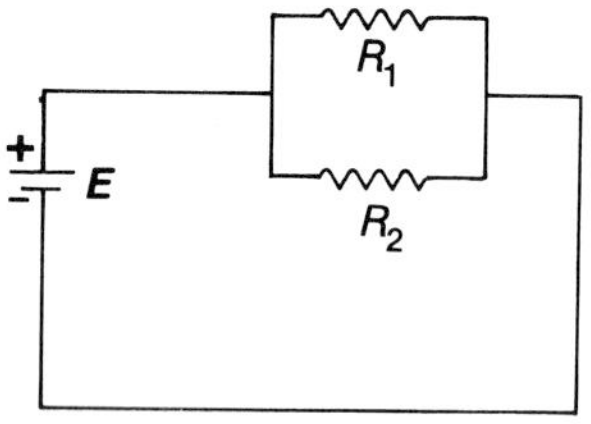

Fig. 4.12. Parallel circuit

In a series circuit the current flow is equal at all points, and its value will be

$$I = \frac{E}{R_1 + R_2 + R_3 + \ldots}.$$

This equation indicates that all resistances in the series circuit are added together and then divided into the voltage (E) to obtain the current.

In a parallel circuit, the total current equals the sum of the currents in the separate branches, and the voltage is equal in each branch to the voltage in the main part of the circuit. For instance, in figure 4.13 the current passing through R_1 will be

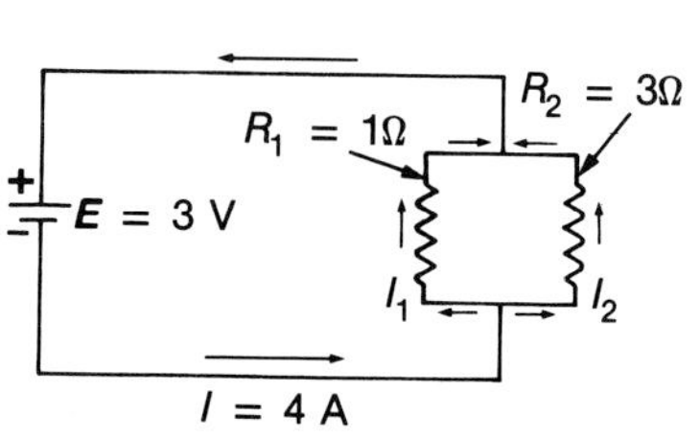

Fig. 4.13 Parallel circuit with total current of 4 amperes

$$I_1 = E/R_1 = 3 \text{ A},$$

while at R_2, it will be

$$I_2 = E/R_2 = 1 \text{ A}.$$

The total current is still 4 amperes.

The effective resistance (R) of a parallel circuit with parallel resistors $R_1, R_2, R_3, \ldots$ is

$$1/R = 1/R_1 + 1/R_2 + 1/R_3 + \ldots$$

For the example above,

$$1/R = 1/1 + 1/3 = 3/3 + 1/3 = 4/3$$
$$R = 3/4, \text{ or } 0.75\Omega.$$

Example Problem: In the circuit in figure 4.14, the initial potential is 12 volts. Calculate the voltage drop across each resistance element.

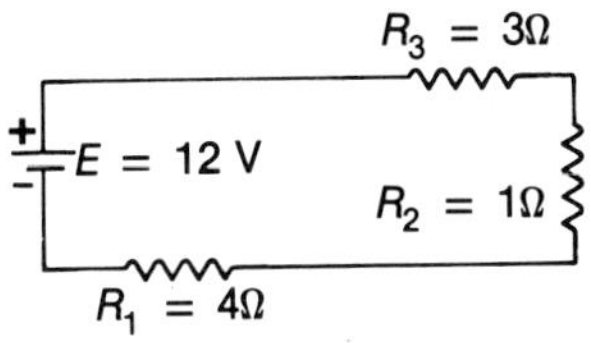

Fig. 4.14

Solution: If the total potential in the circuit is 12 volts, then the total voltage drop for the circuit will be 12 volts. For any resistor, the voltage drop will be proportional to the size of the resistor. The total current in the circuit is

$$I = 12 \div (4 + 1 + 3) = 12 \div 8 = 1.5 \text{ A}.$$

The voltage drop across each resistor is

$$E_1 = 1.5 \text{ A} \times 4 = 6 \text{ V}.$$
$$E_2 = 1.5 \text{ A} \times 1 = 1.5 \text{ V}.$$
$$E_3 = 1.5 \text{ A} \times 3 = 4.5 \text{ V}.$$

The sum of these voltages should equal 12 volts, the initial potential in the circuit:

$$6.0 \text{ V} + 1.5 \text{ V} + 4.5 \text{ V} = 12 \text{ V}.$$

Example Problem: In the parallel circuit in figure 4.15, find the total resistance and the current in each part of the circuit.

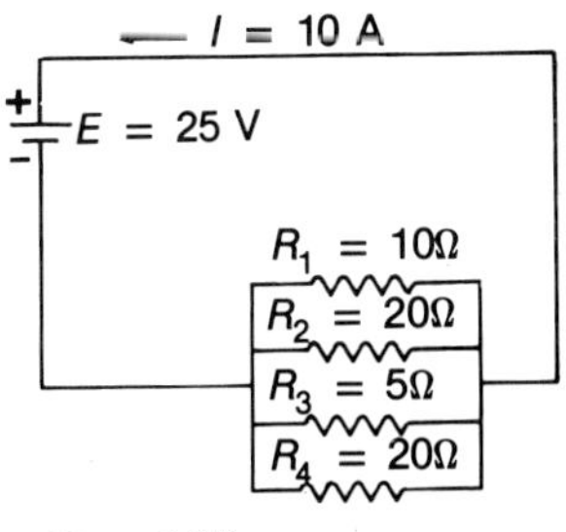

Fig. 4.15

Solution: In each branch of the circuit, the potential is 25 volts. Then at R_1, R_2, R_3, and R_4, the respective currents are

$$I_1 = 25/10 = 2.5 \text{ A}.$$
$$I_2 = 25/20 = 1.25 \text{ A}.$$
$$I_3 = 25/5 = 5.0 \text{ A}.$$
$$I_4 = 25/20 = 1.25 \text{ A}.$$

The total of these currents is seen to be 10 A, the total current in the circuit. The total resistance of the parallel arrangement (R) is

$$1/R = 1/R_1 + 1/R_2 + 1/R_3 + 1/R_4$$
$$= 1/10 + 1/20 + 1/5 + 1/20 = 4/10.$$
$$R = 10/4 = 2.5\Omega.$$

Example Problem: Find the current in each portion of the circuit in figure 4.16 and find the total resistance of the circuit.

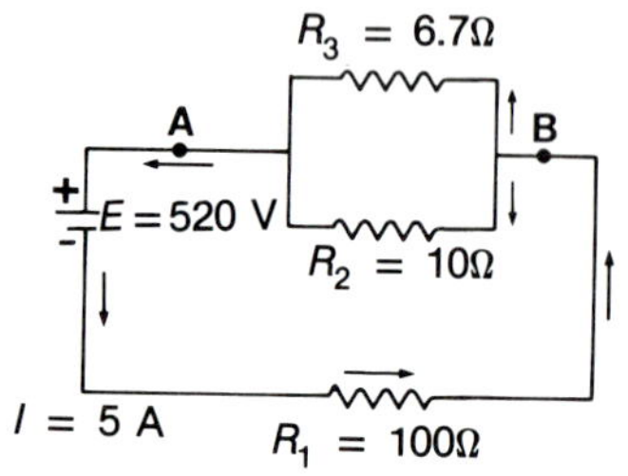

Fig. 4.16

Solution: This is an example of a combination of series and parallel elements in one circuit. The current I at R_1 is 5 amperes. Then the voltage drop at R_1 is

$E_1 = R_1 \times I = 100 \times 5 = 500$ V.

At point B,

$I = 5$ A

$E_B = 520 \text{ V} - 500 \text{ V} = 20$ V.

The current through R_2 is I_2:

$I_2 = E_B/R_2 = 20/10 = 2$ A.

The current through R_3 is I_3:

$I_3 = E_B/R_3 = 20/6.7 = 3$ A.

Effective resistance R of the parallel portion is

$1/R = 1/10 + 1/6.7 = 1.4 = 4\Omega$.

POWER

Power is the rate of doing work or otherwise expending some form of physical energy. In electrical calculations the unit of power is the *watt* (W). In the metric system the watt is the unit of all physical power, both mechanical and electrical. Large amounts of power are expressed in kilowatts (kW) or megawatts (MW). Small amounts of power are typically expressed as milliwatts (mW). Power is related to the three quantities in Ohm's law—voltage, current, and resistance:

$$P = E \times I$$

where

P = power in watts
E = voltage in volts
I = current in amperes.

Since

$$I = E/R,$$
$$P = E \times E/R, \text{ or } P = E^2/R, \text{ and}$$
$$E = I \times R,$$

then

$$P = I \times R \times I, \text{ or } P = I^2R.$$

Example Problem: How much current does a 100-watt, 120-volt bulb draw?

Solution: If $P = E \times I$, then

$I = P/E$
$I = 100/120 = 0.833$ A.

As an energy source, electric power is sold by the kilowatt-hour (kWh). A kilowatt-hour is the amount of work done when 1,000 watts are applied for 1 hour (1,000 watts = 1 kilowatt). The number of kilowatt-hours equals the power (P) in kilowatts times the time (T) in hours:

$$P\,(\text{kW}) \times T(\text{hr}) = \text{kWh}.$$

Example Problem: At 6.2 cents per kilowatt-hour, what does it cost to burn a 100-watt bulb for 100 hours?

Solution: Find the number of kilowatt-hours:

100 W = 0.1 kW.
0.1 kW × 100 hr = 10 kWh.

Then at 6.2 cents per kilowatt-hour, the cost would be

6.2 × 10 = 62 cents.

PRACTICE PROBLEMS

1. If a 12-volt battery supplies current to a circuit having 100-ohm resistance, what is the resultant current? ____________

2. Using the circuit shown, solve the problems that follow.

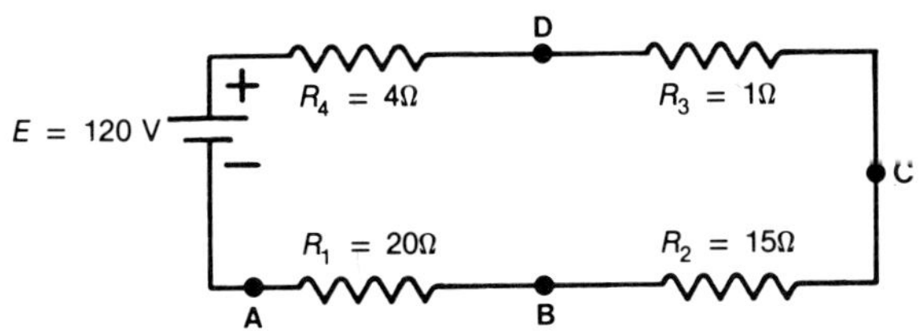

a. What will be the voltage drop across R_1? ____________

b. What is the current at point C? ____________

c. What is the voltage drop across R_4? ____________

d. What is the total current in the circuit? ____________

3. An air conditioner draws 10 amps from a 115-volt circuit for 6 hours per day over a period of 25 days. How much energy in kilowatt-hours is used? ____________

4. Using the circuit below, solve the following problems:

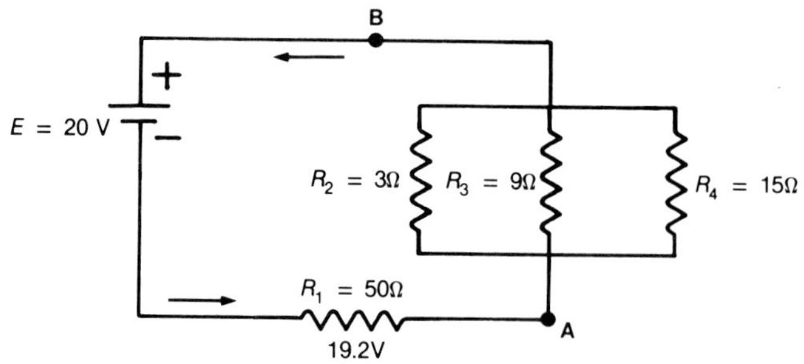

a. What is the effective resistance of the portion of the circuit containing R_2, R_3, and R_4? ______

b. What is the total current in the circuit? ______

c. What is the current through R_2? ______

d. What is the voltage drop across R_1? ______

e. How large a resistor could be added at B to make the current in the circuit equal 0.1 ampere? ______

5. How much power is absorbed by a 20-ohm resistor when it is connected across a 60-volt supply? ______

6. How much energy in watt-hours is delivered by a 6-volt storage battery supplying an average current of 5 amps for an 8-hour period? ______

7. What is the effective resistance of a 6-ohm and a 12-ohm resistor connected in parallel? ______

8. In the figure below, R_2 is rated at 60 ohms.

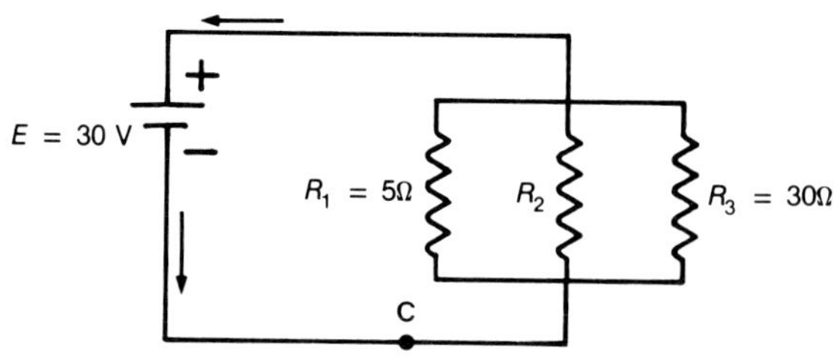

a. What will be the value of the current at point C in the circuit? ______

b. What will be the total power consumed in the circuit? ______

9. How much current does a 60-watt, 120-volt light bulb draw? ______

10. What is the power of an electrical source that has a current of 10 amperes and a resistance of 5 ohms? ______

SELF-TEST

4. Some Physical Quantities and Their Measurement

Multiply each problem answered correctly by five to arrive at your percentage of competency.

Refer to all tables in the chapter to answer these questions.

1. How many miles are in a 10,000-metre race? ____________________

2. How many cubic metres are in a 500-barrel tank? ____________________

3. What is a quantity that cannot be divided into other entities called? ____________________

4. Give the appropriate abbreviation or symbol for the following units of measure:

   a. mile ____________________

   b. pounds per gallon ____________________

   c. cubic foot ____________________

   d. square metre ____________________

   e. pounds per square inch ____________________

   f. grams per cubic centimetre ____________________

   g. pound-force ____________________

   h. ohm ____________________

   i. feet per second ____________________

   j. revolutions per minute ____________________

   k. kilowatt-hour ____________________

5. Give the numerical value of the following metric prefixes:

a. kilo ______________________

b. deci ______________________

c. milli ______________________

d. micro ______________________

e. mega ______________________

6. Convert the following measurements:

a. 64 pounds into kilograms ______________________

b. 3.25 tonnes into tons ______________________

c. 80 newtons into pounds-force ______________________

d. 75 kilograms into pounds-mass ______________________

e. 63 grams into ounces ______________________

7. An electrical capacitor has a capacity of 25×10^{-12} farad. Give the proper expression for that value using the metric prefix and symbol.

____

8. A 30-foot length of pipe weighs 8 pounds per foot. What is the weight of the pipe per metre?

____

9. How many megajoules are in 100 horsepower-hours?

____

10. Convert the following temperatures:

a. 70°F to Celsius ______________________

b. 324K to Celsius ______________________

c. −28°C to Fahrenheit ______________________

d. 203K to Rankine ______________________

e. 152°F to Rankine ______________________

11. A cubic foot of a certain grade of iron weighs 470 pounds. What is its specific gravity?

____

12. The volume of a storage tank filled with fresh water is 5,360 cubic feet. What is the weight of the water in the tank?

____

13. How many cubic feet are in a ton of aluminum?

____

14. A truck winch line pulling with a force of 1,800 pounds pulls an engine 12 feet into position. How much work is being done?

____

15. A diesel engine has 375 horsepower. How many kilowatts of power does it produce?

____

16. A 1,200-watt heating unit is rated at 120 volts. What is its current rating?

____

17. If resistor R_2 in the circuit below is rated at 40 ohms, what will be the voltage drop across each resistor?

____

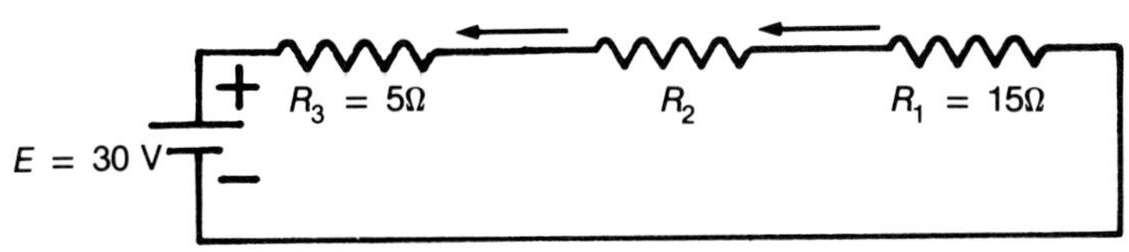

18. What is the pressure at the bottom of a 1,600-barrel water storage tank when the height of the water is 16 feet?

____

19. Convert the following temperatures and pressures to *absolute* values:

a. 110°F ______________________

b. 23°C ______________________

c. 39 psig ____________________

d. 16 psig ____________________

e. 1,300°F ____________________

20. A tank of gas at 99°F had a pressure of 25 psig. When the pressure dropped to 10 psig, what was its temperature?

____

5
PRINCIPLES OF ALGEBRA

Objectives

Upon completion of chapter five, the student will be able to—

1. Use *variables* to represent an unknown quantity.
2. Solve an equation for an unknown quantity.
3. Convert a written problem into an equation and solve the equation.
4. Simplify an expression by removing the grouping symbols.
5. Add two or more algebraic expressions.
6. Add algebraic expressions containing a negative quantity.
7. Subtract one algebraic expression from another.
8. List the rules for multiplying algebraic expressions.
9. Multiply algebraic expressions containing exponents.
10. Divide one algebraic expression by another.
11. Transpose an equation to solve for a variable.
12. Use formulas to solve for quantities often used in industrial applications.

In simple arithmetic, figures and signs are used exclusively in making calculations. In algebra, letters are also used. The letters are called *variables* and are used to denote unknown quantities in algebraic expressions. A variable may denote different values in different problems, but it always represents the same value in a given problem. That is, x may stand for gallons in one problem and for feet in another, but, if it is selected to stand for gallons in a problem, it must represent gallons throughout that problem.

Algebraic expressions—combinations of symbols, letters, and signs—are used in formulas and equations. A *formula* is a symbolic expression of a general fact, rule, or principle and is often stated as an equation. An *equation* is an expression of equality between two quantities.

Algebraic Expressions

An algebraic expression is a combination of algebraic symbols and signs, such as $2a + 3ab - 4c$. The *terms* of the expression are the combinations of symbols that are not separated by a sign; for example, $2a$, $3ab$, and $4c$ are the terms of the expression $2a + 3ab - 4c$. Algebraic expressions may be added, subtracted, multiplied, and divided. Grouping terms indicates the order in which these operations should be carried out.

GROUPING SYMBOLS

Parentheses (), brackets [], and braces {} may be used for grouping terms. These symbols are used to indicate that certain terms should be kept together and to indicate the order in which various operations should be performed. These grouping symbols must be removed before addition or subtraction can be performed. A few rules govern the use of grouping symbols.

Rule 1. When an expression within parentheses is preceded by the plus sign, the parentheses may be removed without changing the signs within the parentheses.

Rule 2. When an expression within parentheses is preceded by the minus sign, the parentheses may be removed, provided the signs within the parentheses are changed.

Rule 3. A number or symbol before parentheses indicates that each term within the parentheses is to be multiplied by that number or symbol.

Example Problem: Simplify the expression, $3(a - b) + 4(b + c)$.

Solution: First remove the parentheses:

$$3a - 3b + 4b + 4c.$$

Then combine like terms:

$$3a + b + 4c.$$

Example Problem: Simplify the following expression by removing the symbols of grouping and by combining like terms:

$$3x + 4[2x - 3(x - 7y) + (x + 3y)]$$

Solution:

$$\begin{aligned} &3x + 4(2x - 3x + 21y + x + 3y) \\ &3x + 8x - 12x + 84y + 4x + 12y \\ &3x + 96y \end{aligned}$$

ADDITION

Addition in algebra consists of collecting the terms of an expression into an equivalent expression. *Like terms* (terms that are alike) are placed in vertical columns. For example, to add $2a + 3b + c$ and $a + 6b + 8c$, the terms should be arranged in this manner:

$$\begin{array}{r} 2a + 3b + c \\ a + 6b + 8c \\ \hline 3a + 9b + 9c \end{array}$$

Both positive and negative quantities may be added. Adding a negative quantity (a quantity with a minus sign before it) is equivalent to subtracting a positive quantity of the same value. That is, $-3z$ added to $5z$ is equal to $2z$. Adding negative quantities may also be expressed as $5z + (-3z) = 2z$.

Example Problem: Add $3a + 2ab - c$ and $3ab + 4c$.

Solution:

$$\begin{array}{r} 3a + 2ab - c \\ 3ab + 4c \\ \hline 3a + 5ab + 3c \end{array}$$

Example Problem: Add $3x + 2xy - 3y$ and $5x - 3xy + y$ and $3x + 4y$.

Solution: Arrange like terms in vertical columns and add:

$$\begin{array}{lcr} 3x & + 2xy & - 3y \\ 5x & - 3xy & + y \\ 3x & & + 4y \\ \hline 11x & - xy & + 2y \end{array}$$

SUBTRACTION

Subtraction may be considered as the inverse operation of addition. Subtraction is carried out by changing the sign (+ becomes −; − becomes +) of each term in the subtrahend and then proceeding as in addition. Thus, $3x^2 + 5x + 10$ minus $x^2 + 2x - 4$ would be carried out as follows:

$$\begin{array}{r} 3x^2 + 5x + 10 \\ -x^2 - 2x + 4 \\ \hline 2x^2 + 3x + 14 \end{array}$$

If a term appears in the subtrahend and no like term appears in the minuend, the term in the subtrahend is brought down into the result with its sign changed. For example, $3a - 4ab + 3c$ would be subtracted from $6a + 4c$ as follows:

$$\begin{array}{rrr} 6a & & + 4c \\ -(3a & - 4ab & + 3c) \\ \hline 3a & + 4ab & + c \end{array}$$

Example Problem: Subtract $2x^2 + 3x - 1$ from $5x^2 + x + 3$.

Solution:

$$\begin{array}{rrr} 5x^2 & + x & + 3 \\ -(2x^2 & + 3x & - 1) \\ \hline 3x^2 & - 2x & + 4 \end{array}$$

Example Problem: Subtract $6a^2 - 3a + 4ab$ from $12a^2 - b^2 + 3ab$.

Solution

$$\begin{array}{rrrr} 12a^2 & & + 3ab & - b^2 \\ -(\ 6a^2 & - 3a & + 4ab &) \\ \hline 6a^2 & + 3a & - ab & - b^2 \end{array}$$

MULTIPLICATION

Multiplication may be indicated by several symbols:

$a \times b = c.$

$a \cdot b = c.$

$ab = c.$

$a(b) = c.$

Four simple rules govern algebraic multiplication.

Rule 1. A positive term multiplied by a positive term results in a positive product:

$$3a \times b = 3ab.$$

Rule 2. A negative term multiplied by a negative term results in a positive product:

$$-3a \times (-b) = 3ab.$$

Rule 3. A positive term multiplied by a negative term results in a negative product:

$$3a \times (-b) = -3ab.$$

Rule 4. When like quantities are multiplied, their exponents are added in the product (when no exponent is present, it is considered to be 1):

$$3a \times a^2 = 3a^{2+1} = 3a^3.$$

$$4b^2 \times 2b^3 = 8b^{2+3} = 8b^5.$$

Example Problem: Multiply $3a - 2b$ by $3b$.

Solution:

$$\begin{array}{rcr} 3a & - & 2b \\ \times & & 3b \\ \hline 9ab & - & 6b^2 \end{array}$$

To multiply expressions when there are two or more terms in the multiplier, follow these steps:

1. Multiply each term in the multiplicand by the last term.
2. Multiply each term in the multiplicand by the next to the last term and all other terms, working from right to left and placing like terms under each other. Unlike terms go in a separate column.
3. Then add like terms.

Example Problem: Multiply $-8a + 6b + 4c$ by $b + c$.

Solution: Multiply each term in the multiplicand by c, then by b, and add like terms:

$$\begin{array}{lrrrrl} & -8a & + 6b & + 4c & & \\ \times & & b & + c & & \\ \hline & -8ac & + 6bc & + 4c^2 & & \text{(Step 1)} \\ & & 4bc & - & -8ab + 6b^2 & \text{(Step 2)} \\ \hline & -8ac & + 10bc & + 4c^2 & -8ab + 6b^2 & \text{(Step 3)} \end{array}$$

DIVISION

Division of algebraic expressions involves the same processes of multiplication and subtraction that are used in solving arithmetic problems. The following rules apply to division of algebraic expressions.

Rule 1. When the signs of the divisor and the dividend are alike, the quotient is positive:

$$-10x \div (-2x) = 5x.$$

Rule 2. When the signs of the divisor and dividend are different, the quotient is negative:

$$4a \div (-2a) = -2a.$$

Rule 3. When a term is divided by a like term, the exponent of the divisor is subtracted from the exponent of the dividend:

$$x^5 \div x^3 = x^{5-3} = x^2.$$

Example Problem: Divide $6x^2 + 9xy - 6x$ by $3x$.

Solution: Write down the dividend and divisor in the following manner:

$$3x \,\overline{)\, 6x^2 + 9xy - 6x}$$

Divide each term of the dividend by the divisor; write down each quotient; place the appropriate sign before each quotient.

$$\begin{array}{r l}
 & 2x \quad + 3y \quad - 2 \\
3x & \overline{)\, 6x^2 + 9xy - 6x} \\
 & \underline{6x^2} \\
 & \qquad\quad 9xy \\
 & \qquad\quad \underline{9xy} \\
 & \qquad\qquad\qquad 6x \\
 & \qquad\qquad\qquad \underline{6x}
\end{array}$$

Example Problem: Divide $x^2 - 2xy + y^2$ by $x - y$.

Solution: Arrange the dividend and divisor as for long division of numbers in arithmetic.

$$\begin{array}{r l}
 & \qquad\quad x \quad - y \\
x - y & \overline{)\, x^2 - 2xy + y^2} \\
 & \underline{x^2 - \;\; xy} \\
 & \qquad\quad xy + y^2 \\
 & \qquad\quad \underline{xy + y^2}
\end{array}$$

Example Problem: Divide $a^4 + a^3 - 9a^2 - 16a - 4$ by $a^2 + 4a + 4$.

Solution: Arrange the dividend and divisor as in the previous example.

$$\begin{array}{r l}
 & \qquad\qquad\qquad a^2 - 3a - 1 \\
a^2 + 4a + 4 & \overline{)\, a^4 + a^3 - 9a^2 - 16a - 4} \\
 & a^4 + 4a^3 + 4a^2 \\
 & \quad - 3a^3 - 13a^2 - 16a \\
 & \quad \underline{- 3a^3 - 12a^2 - 12a} \\
 & \qquad\qquad\quad - a^2 - 4a - 4 \\
 & \qquad\qquad\quad \underline{- a^2 - 4a - 4}
\end{array}$$

PRACTICE PROBLEMS

1. Find the sum of the following: $2a + 3b$ and $a - 6b$. __________
2. Add $3b - 4bc$ and $5a + 3bc + 2c$ and $6a + 4c$. __________
3. Find the sum of $6a + 20b + 4ab + c$ and $6b - 8ab - 9c$ and $-4a + 6c$. __________
4. Add $5x + 3xy + 4y$ and $-8x - 2xy$ and $-3y$. __________
5. Find the sum of $-8ab + 2a + c$ and $7a + 9ab - 4c$ and $8c - 12ab + 6a$ and $8b + 4c + 3ab$. __________
6. Subtract $3a^2 - 2b + c$ from $4a^2 - 6b + 2c$. __________
7. Take $6x + 3y$ from $4x^2 - 6xy - 4y$. __________
8. Subtract $6a(b + c) + 4(a + b)$ from $9a(b - c) - 8(b + c)$. __________
9. Subtract $4a^2 + 3ab + 6c^2$ from $9a^2 - 8c^2$. __________
10. Subtract $4(x + xy) - 6y$ from $6x^2 - 6(x + 2y)$. __________
11. Multiply $x^2 + 2xy + y^2$ by $4a$. __________
12. Multiply $4a + 6ab + 3c$ by $5ab$. __________
13. Multiply $3a + 4b$ by $a - b$. __________
14. Multiply $(a + 3b) - (a + b)^2$ by $4a$. __________
15. Simplify the expression $(x + y)^2 - (2x - 3y)$. __________
16. Divide $8a^2 + 4ab$ by $-2a$. __________
17. Divide $27a^3 + 9a^2 + 12ab + 6b$ by $3a$. __________
18. Divide $m^4 + 57m - 70$ by $m^2 + 3m - 5$. __________
19. Divide $x^3 - 3x^2y + 3xy^2 - y^3$ by $x - y$. __________
20. Divide $m^4 + 64$ by $m^2 + 4m + 8$. __________

Equations

An *equation* may be simple or it may be quite complicated, but it is always an expression of two quantities that equal each other. For example, $2 = 2$ is a simple equation. Another simple equation is $3 \times 4 = 2 \times 6$. If variables are used, the equation might read

$$3a = 2b$$

where

$a = 4$
$b = 6.$

Example Problem: If $a = 6$, $b = 4$, and $c = 3$, find the value of $2a + 3b + 5c$.

Solution: Substituting numerical values for symbols,

$$(2 \times 6) + (3 \times 4) + (5 \times 3) = 12 + 12 + 15 = 39.$$

The two sides of an equation (left and right of the equals sign) must remain equal, no matter what operation is performed to solve for the variable. If an arithmetical operation is performed on one side of the equation, the same operation must be performed on the other side of the equation.

SOLVING PROBLEMS WITH EQUATIONS

When the numerical value of a quantity is unknown, it is represented by a letter or algebraic symbol. Then, using the data given, an equation is set up. If enough data are given, the numerical value represented by the letter can be obtained. Following are the general steps for solving problems involving simple equations:

1. Let a letter or symbol represent one unknown number.
2. If there is a second number, represent it in terms of the first unknown, using the conditions given in the problem.
3. Set up an equation, using the relations given by the problem.
4. Solve for the unknown number.
5. Use the new number to solve for other unknown numbers, using the conditions given in the problem.
6. Check the numbers to see whether they meet the conditions of the problem.

Example Problem: The sum of two numbers is 92, and their difference is 4. Find the numbers.

Solution:

Let x = the greater number, and $x - 4$ = the lesser number. Set up the equation according to the conditions of the problem:

$$x + x - 4 = 92.$$

Solve for x:

$$2x - 4 = 92$$
$$2x = 92 + 4 = 96$$
$$x = 96 \div 2 = 48, \text{ the greater number.}$$

Then solve for the other unknown:

$$48 - 4 = 44, \text{ the lesser number.}$$

Finally, check to see whether these numbers meet the conditions in the problem:

$$48 + 44 = 92.$$
$$48 - 44 = 4.$$

Example Problem: Ten steel plates of a certain size weigh 60 pounds each. Their total weight is equal to the weight of eight plates of a different size. What is the weight of one of the eight plates?

Solution: Choose some symbol such as y to equal the weight of one of the eight plates.

10×60 lb = 600 lb, the weight of the ten plates.
$8y$ = the weight of the other eight plates.

But the total weights are the same, 600 lb; therefore,

$8y = 600$ lb and $y = 600/8 = 75$ lb, the weight of one plate.

PRACTICE PROBLEMS

1. Let $a = 12$, $b = 4$, and $c = 2$. Find the value of the following:

   a. $2a + 3b$ ________________

   b. $a + 2b + c$ ________________

   c. $3b - c$ ________________

2. In the equation $a + 2b + c = 14$, what is the value of a if $b = 2$ and $c = 8$? ________________

3. Write the following equations, using the symbols a, b, c, d, and e where $a = 4$, $b = 1$, $c = 5$, $d = 2$, and $e = 3$.

   a. $4 + 5 + 1 = 3 + 4 + 1 + 2$ ________________

   b. $1 + 5 + 3 = 3 + 1 + 2 + 3$ ________________

   c. $5 + 4 + 2 = 5 + 3 + 1 + 2$ ________________

4. A man receives \$572 for 8 days of work. Using an algebraic equation, determine how much he receives per day. ________________

5. Five pounds of copper can be purchased at \$1.07 per pound. An inferior grade may be bought for 72¢ per pound. By means of an equation, determine how many pounds of the inferior grade could be purchased for the same money that would have to be expended for 5 pounds of the better grade. ________________

6. Divide 35 into two parts, so that 4 times the lesser number equals 3 times the greater number.

   Greater number: ________________

   Lesser number: ________________

7. A hundred and five feet of chain link fence material is available to build a square pen. An existing wall will form one side of the pen. What is the maximum area that can be enclosed by the wall and the fence material? ________________

8. A small ship is chartered for a weekend excursion for 100 passengers at \$150 each. The ship's owner agrees to reduce the rate \$10 per person for each additional passenger. What is the total cost per person if 130 people participate in the cruise? ________________

9. A student received *A* grades on 50% of one term's assignments and *B* grades on five other assignments for the term. The remaining term assignments, or 4/9 of the total, were not handed in. How many assignments were there? ________________

10. A speculator bought 100 shares of stock. He sold 25 shares at twice their cost, then the company went into bankruptcy. The remaining shares were sold for 20¢ each. His net loss was \$235. What was the price per share initially? ________________

Formulas

A *formula* consists of a group of numerical symbols arranged in such a way as to express briefly a single concept. A formula may be a simple term or expression, or it may be in the form of an equation. Formulas are used to save space and time when writing and working problems. For example:

$$T_m = D \times T_d$$

where

T_m = money earned in one month
D = number of days worked
T_d = money earned in one day.

Instead of using the formula, the same information can be written: "The total amount of money earned in one month is equal to the number of days worked that month multiplied by the amount of money earned in one day." The formula is obviously easier to use. In all formulas, when letters and symbols are used in place of numbers, *the letters or symbols must be defined.*

Example Problem: A pumper earns $1,625 per month and works 12 months per year. His average living expenses are $1,345 per month. The balance is placed in a savings account. How much does he save in a year?

Solution: To reduce the statement to an algebraic formula, choose a letter for each factor to be considered in solving the problem. Let D stand for the number of dollars saved in a year, S for monthly earnings in dollars, and E for monthly expenses in dollars. Now using these letters, set up the algebraic formula:

$$D = (12 \times S) - (12 \times E)$$

or written in a simpler form,

$$D = 12S - 12E$$

where

D = number of dollars saved in a year
S = monthly earnings in dollars
E = monthly expenses in dollars.

Then solve for D by substituting figures for the known factors:

S = \$1,625; E = \$1,345.
$D = (12 \times 1{,}625) - (12 \times 1{,}345)$ = \$3,360 saved in a year.

TRANSPOSITION

Sometimes it is necessary to change formulas around, or *transpose* the terms, in order to find the unknown quantity. For example, consider the formula

$$T_m = D \times T_d$$

where

T_m = money earned in one month
D = number of days worked
T_d = money earned in one day.

Suppose the amount of the monthly check and the number of days worked are known, and the daily rate of pay is needed. The formula would then be written to solve for T_d, the money earned in one day:

$$T_d = T_m \div D.$$

Note that transposing terms of an equation from one side of the equals sign to the other changes the operational sign. In this case the sign changes from multiplication to division. That is, the daily rate is equal to the monthly check divided by the number of days worked during the month. The formula should always start with the unknown quantity; however, it can be reversed for calculator purposes once it is established.

A formula that is probably the most commonly used one in business and industry today is $a/b = c/d$ and its ramifications. It is the basis for setting up proportions and solving for an unknown when three factors are known. Substituting numbers for the variables, $a = 12$, $b = 6$, $c = 4$, and $d = 2$:

$$12 \div 6 = 4 \div 2,$$

or

$$12/6 = 4/2.$$

Since both terms are equal to 2, they are equal to each other and constitute an equation.

Transposing terms in this equation results in the formula for direct proportion:

$$ad = bc,$$

which comes from the statement, *the product of the means equals the product of the extremes.* In a direct proportion, the first and fourth terms (a and d) are called the extremes, and the second and third terms (b and c) are called the means. The four transpositions of this formula are commonly used:

$a = bc/d$ (both sides divided by d)
$b = ad/c$ (both sides divided by c)
$c = ad/b$ (both sides divided by b)
$d = bc/a$ (both sides divided by a)

Example Problem: A photograph that is 8 inches wide by 10 inches deep is going to be reduced to 3 inches wide to fit a magazine column. How deep will the reduced picture be?

Solution: Set up the proportion by saying, "8 is to 10 as 3 is to *x*." Since the fourth term d is the unknown x, use the formula to solve for d and substitute the known numbers for the letters: $a = 8$, $b = 10$, and $c = 3$.

$$d = bc/a$$
$$d = 10 \times 3 \div 8 = 3.75''.$$

Transposing terms and expressions of an equation can involve several steps. The important rule to remember is that the equation remains valid as long as each side is multiplied by, divided by, added to, or subtracted from equally. Of course, multiplication and division by the same term on the same side of the equation cancels that term, and addition and subtraction of the same term on the same side of an equation cancels the term. To illustrate, solve for b in the equation

$$a = 20 \div (b + 10).$$

First, multiply each side by the expression $(b + 10)$:

$$a(b + 10) = 20 \div (b + 10) \times (b + 10).$$

The last two terms cancel each other, leaving

$$a(b + 10) = 20.$$

Then, divide both sides by a:

$$b + 10 = 20/a.$$

The last step is to subtract 10 from each side:

$$b = 20/a - 10.$$

FORMULA APPLICATIONS

Formulas are used extensively in the petroleum and other industries. Many of the commonly used formulas deal with area and volume and therefore are given in chapter six, "Practical Geometry." Other established formulas and their applications in industry are presented here and in chapter eight.

Since most calculations are carried out with a calculator, formulas can be written in a sequence that will give the desired answer after the numbers and operation signs have been entered in the proper sequence. Most formulas *begin* with the unknown factor; therefore, they must be restated or rewritten in a sequence that *ends* with the unknown or desired factor. For example,

$$a = \frac{b + c}{d}$$

should be restated to read

$$(b + c) \div d = a.$$

Well control formulas. To calculate hydrostatic pressure, mud weight, and true vertical depth, the following formula and its derivatives are used:

$$C \times MW \times TVD = P$$

where

C = constant of 0.052 for pounds per gallon (ppg); 0.007 for pounds per cubic foot (pcf)
MW = mud weight in ppg or pcf
TVD = true vertical depth in feet
P = hydrostatic pressure in pounds per square inch (psi).

Transposing this formula,

$$\frac{P}{TVD \times C} = MW$$

and

$$\frac{P}{MW \times C} = TVD.$$

Example Problem: What is the hydrostatic pressure in a well whose mud weight is 11.2 ppg and whose true vertical depth is 4,000 feet? As stated, the constant for ppg is 0.052.

Solution:

$C \times MW \times TVD = P$
$0.052 \times 11.2 \times 4{,}000 = P$
2,329.6 psi $= P$ (hydrostatic pressure).

To perform kill mud weight calculations, the following formulas are used:

$$\frac{SIDPP}{TVD \times C} = MWI,$$

$$\frac{P_O}{TVD \times C} = MWI_O$$

and

$$OMW + MWI + MWI_O = KMW$$

where

$SIDPP$ = shut-in drill pipe pressure
TVD = true vertical depth
C = constant (0.052 for ppg)
MWI = mud weight increase required to compensate for surface pressure on the drill pipe (shut-in drill pipe pressure)
P_O = overbalance pressure
OMW = original mud weight
MWI_O = mud weight increase required for overbalance
KMW = kill mud weight.

Example Problem: Given the following factors, find the mud weight increase needed to compensate for the surface pressure on the drill pipe:

Shut-in drill-pipe pressure = 310 psi
True vertical depth = 7,000 ft
Constant for ppg = 0.052
Original mud weight = 9.5 ppg
Overbalance pressure = 200 psi.

Solution:

$$SIDPP \div (TVD \times C) = MWI$$

or simplified for the calculator:

$$SIDPP \div TVD \div C = MWI.$$
$$310 \div 7{,}000 \div 0.052 = 0.8516.$$

Rounded off, mud weight increase = 0.9 ppg.

Instrumentation and electrical formulas. In the study of instrumentation, electricity, and electronics, many formulas are used to calculate variables. These formulas are usually based on proven facts or laws of physics. For example, the ideal gas law, which combines Boyle's law and Charles's law, states that *the volume of a quantity of gas varies inversely as the absolute pressure and directly as the absolute temperature. Expressed as a formula,*

$$P_1 V_1 \div P_2 V_2 = T_1 \div T_2$$

where

P_1 = original pressure
P_2 = final pressure
V_1 = original volume
V_2 = final volume
T_1 = original temperature
T_2 = final temperature.

Pressures and temperatures in the formula are *absolute.* To convert gauge pressure to absolute pressure, add 14.7 psi to the gauge pressure. To convert Fahrenheit temperature to Rankine (absolute) temperature, add 460° to the Fahrenheit temperature. To convert Celsius temperature to Kelvin (absolute) temperature, add 273° to the Celsius temperature.

Example Problem: A volume of gas fills 1,000 cubic feet at 80 pounds per square inch gauge (psig) and 80°F. What would be the volume of the gas at 100 psig and 60°F?

Solution: Transpose the formula:

$V_2 = P_1 V_1 T_2 \div P_2 T_1$.

Substitute figures for the factors:

P_1 = 80 psig + 14.7 psi = 94.7 psia
P_2 = 100 psig + 14.7 psi = 114.7 psia
T_1 = 80°F + 460° = 540°R
T_2 = 60°F + 460° = 520°R
V_1 = 1,000 cubic feet

Then

$$V_2 = (94.7 \times 1{,}000 \times 520) \div (114.7 \times 540),$$

or restated for the calculator:

$$94.7 \times 1{,}000 \times 520 \div 114.7 \div 540 = 795.05$$

So, at 100 psig and 60°F, the volume of gas would be 795 cubic feet.

To change temperature from degrees Celsius to degrees Fahrenheit:

$$°F = \frac{9}{5}°C + 32°$$

where

°F = temperature in degrees Fahrenheit
°C = temperature in degrees Celsuis.

Likewise, to change temperature from degrees Fahrenheit to degrees Celsius, the formula is

$$°C = \frac{5}{9}(°F - 32°).$$

In electrical calculations, the basis of many formulas is Ohm's law, which states: *The current flowing in a circuit, in amperes, is directly proportional to the electromotive force (emf), in volts, and inversely proportional to the resistance of the circuit, in ohms.* Expressed as a formula, this statement becomes

$$I = E/R$$

where

I = the current in amperes
E = emf in volts
R = resistance in ohms.

The equation may be transposed to solve for the resistance if the current and voltage are known:

$$R = E/I;$$

or to solve for the voltage if the current and resistance are known:

$$E = I \times R.$$

Unit conversions. Many formulas are involved in converting from conventional units of measure to metric units and vice versa. For example, to convert miles to kilometres, the number of miles is multiplied by a conversion factor of 1.609 (as shown in table 4.3). To set up a formula for conversion:

$$km = mi \times C$$

where

km = kilometres
mi = miles
C = conversion factor of 1.609.

Since determining the multiplier constant in such calculations often becomes involved, tables have been established for quick conversion of units (see tables 4.2, 6.1, and 6.3). Even with tables, some calculating is necessary to get the answer in the appropriate unit of measure.

Example Problem: If 1 cubic foot is equivalent to 7.48 gallons, what capacity tank would be needed to hold 50 gallons?

Solution: Set up a proportional equation, letting x equal the tank capacity needed to hold 50 gallons:

$1 : 7.48 = x : 50$
$x = 1 \times 50 \div 7.48$
$x = 6.684 \text{ ft}^3$

PRACTICE PROBLEMS

1. The formula for finding the permeability of a formation sample is

$$C = \frac{\mu QL}{A(P_1 - P_2)}$$

where

C = permeability, in darcys
μ = viscosity of fluid used, in centipoises
Q = volume of water passing through sample, in cubic centimetres per second
P_1 = input pressure, in atmospheres
P_2 = delivery pressure of fluid leaving sample, in atmospheres
L = length of sample, in centimetres
A = area of cross section of sample, in square centimetres.

Given the following values for the variables, what will be the permeability of the sample?

$\mu = 15$
$Q = 0.5$
$P_1 = 4$
$P_2 = 1$
$L = 12$
$A = 75$

(Note: The units will cancel out so that the answer will be in darcys—the usual unit of measure used to describe permeability.) __________

2. Convert the following metric measurements to conventional units: (Refer to table 4.3 for conversion factors.)

   a. 6 millimetres to inches __________

   b. 152 kilometres to miles __________

   c. 6 metres to yards __________

   d. 30 centimetres to inches __________

   e. 2 metres to feet __________

3. If 45 floor tiles cost $32.50, how much will 180 tiles cost? __________

4. If a litre of gasoline sells for 35.9 cents per litre, how many gallons can you buy for $5.00? (1 gallon = 3.785 litres) __________

5. Set up a formula to solve for each of the following variables of an automobile trip if the other two variables are known:

   a. The distance traveled (D), in miles __________

   b. The average speed (S), in miles per hour __________

   c. The time required for the trip (T), in hours __________

6. A formula useful in estimating the approximate number of barrels of oil in a tank is

$$V = d^2 \times 0.14 \times h$$

where

V = volume of oil, in barrels
d^2 = diameter of tank, in feet
0.14 = a constant
h = height of fluid, in feet.

Using this formula, find the approximate number of barrels of oil in a tank when the height of the oil is 30 feet and the tank diameter is 114.5 feet. ____________

7. Transpose the following formula for finding °API gravity (API) to solve for specific gravity (SG):

$$API = 141.5 \div (SG - 131.5).$$

SG = ____________

$API = \frac{141.5}{S.G.} - 131.5$

8. Using the formula in problem 7, find the °API gravity of a crude oil whose specific gravity is 0.8311. ____________

9. Boyle's law expressed as a formula is

$$P_1 V_1 = P_2 V_2$$

where

P_1 = original pressure of gas (absolute)
V_1 = original volume of gas
P_2 = new pressure of gas (absolute)
V_2 = new volume of gas.

Transpose this equation to solve for each of the terms involved. Show the four formulas.

P_1 = ____________

V_1 = ____________

P_2 = ____________

V_2 = ____________

10. A volume of gas fills 2,000 cubic feet at 80 psig. If the temperature remains constant, what will be the volume of 100 psig? (*Hint:* Convert psig to psia; then use the formula given in problem 9.) ____________

SELF-TEST
5. Principles of Algebra

Multiply each problem answered correctly by five to arrive at your percentage of competency.

1. Solve for x in the following equations:

   a. $3x = 12$ ____________

   b. $24 - x = 15$ ____________

   c. $x + 15 = 45$ ____________

   d. $2(2x - 4) + 4x = 152$ ____________

   e. $32x + 50x = x^2 + 3.5x + 10^3$ ____________

2. If $a = 6$, $b = 4$, and $c = 3$, find the value of each of the following:

   a. $3a + 6b + c$ ____________

   b. $a^2 + ab + 2c$ ____________

   c. $-3a^2 + 2ab - 2c^2$ ____________

   d. $2c^3$ ____________

   e. $3a + 2ab + 5b + bc + c^2$ ____________

3. Ten stands of pipe of a certain diameter weigh 900 pounds each. Their total weight is equal to the weight of six stands of pipe of a different diameter. What is the weight of one of the heavier stands of pipe?

   ____

4. Add the following algebraic expressions:

   a. $5x + 3y$ and $4x - 6y$ ____________

   b. $3x + 4y$ and $-4x - 8y$ ____________

   c. $5x - 3y + 4z$ and $8x - 2z + 3y$ and $7x - 5y - 9z$

   ____

d. $3a + 4b - 6c + 8d$ and $7c + 4d$ and $-2a + b$ and $7c + 9d$

____

e. $2x^2 + 4x^2y + 8xy^2$ and $3x^2 + 9xy^2$ and $-4x^2 + 3x^2y$ and $4x^2y - 9xy^2$

____

5. Subtract the following algebraic expressions:

a. $2a - 2b + c$ from $a - 2b - 3c$ ____________________

b. $4a - 5b + 3c$ from $4a - 8b + 2c$ ____________________

c. $2x - y + 5z + 6$ from $3x - 2y + z$ ____________________

d. $4a^2b - 5ab^2 - b^3$ from $3a^2 - 3a^2b - 3ab^2$

____

e. $x^3 - z^3 - 3x^2z + 3xz^2$ from $5x^2z + 8z^3 + 3x^3 + 2xz^2$

____

6. Simplify the following expressions by removing the grouping symbols and combining like terms.

a. $(4a - 3b + 6) + (3a + 5b - 10)$ ____________________

b. $(2x^2 + 3x + 10) + (x^2 - 5x + 5) - (x^2 + x + 3)$

____

c. $5x - [2x + (-3x + 4y) - (x - 8y) + 4x]$

____

d. $(5a - 6b) - [-4a - (4c - b) - 2c]$ ____________________

e. $-\{[(x^2 - a^2) + (y^2 - 2a^2)] - [(x^2 - a^2) - (y^2 - 2a^2)]\}$

____

7. Multiply the following expressions:

a. $-5ab \times b$ ____________________

b. $2x^2 \times 3x^3$ ____________________

c. $-4x^5 \times -6x^3$ __________

d. $(x + y)^2 \times 4xy$ __________

e. $(-x^3ay) \times (x^2 - 2ax + y^2)$ __________

8. Divide the following:

a. $8a^3 \div a^2$ __________

b. $(-18x^7) \div (6x^4)$ __________

c. $15x^2y^4 \div 75xy^4$ __________

d. $\dfrac{(a + x) - 2(a + x)^3}{a + x}$ __________

e. $(a^3 + 5a^2b + 5ab^2 - 3b^3) \div (a + 3b)$ __________

9. An owner sold $2/5$ of his farm to one man and 14 acres to a second man. Find the number of acres in the original farm if the owner had half of the farm left. (*Hint:* Let x equal the total number of acres in the original farm.)

____

10. Using Ohm's law, $I = E/R$, find the resistance (R) of a load if the voltage (E) across the load in a circuit is 120 volts and the current (I) through the load is 6 amperes.

____

11. Solve for the unknown in the following equations:

a. $A = \frac{1}{2}h(b + c)$, where $h = 13$, $b = 7$, $c = 11.2$.

____

b. $h = r + \frac{1}{2}\sqrt{4r^2 - c^2}$, where $c = 6.5$, $r = 8.7$.

____

c. $d = t/3\sqrt{rs}$, where $r = 4.2$, $t = 7$, $s = 5$.

____

d. $A = bc/d + c$, where $c = 6.8$, $d = 2.5$, $b = 9.2$.

____

e. $S = 1.157/P - A$, where $S = 0.04764$, $A = 0.035$.

____

12. Separate 55 into two parts so that two-thirds of the larger part is 5 more than the smaller part.

Larger part: _______________

Smaller part: _______________

13. A car that is 14 feet long overtakes a 28-foot truck that is going 50 miles per hour. How many feet per second must the car travel in order to pass the truck in 4 seconds?

____

14. What is the speed of the car in problem 13, in miles per hour?

____

15. A company has prepared a safety booklet that it will sell for $2.00 each. Under plan A, the company can get 15,000 copies printed for $5,000. Under plan B, it can get twice as many copies for one and a half times the cost.

a. What will be the cost per book under each plan?

Plan A: _______________

Plan B: _______________

b. How many books does the company have to sell before it starts making a profit?

Plan A: _______________

Plan B: _______________

c. After costs are recovered, how many books will be left to sell under each plan?

Plan A: _______________

Plan B: _______________

d. Assuming all books are sold, what is the total profit under each plan?

Plan A: _______________

Plan B: _______________

e. The company sold 25,000 copies and then decided to revise the booklet. Under plan A, the company would have to reorder 10,000 more copies at a printing cost of $4,500. Under plan B the company would have to throw away 5,000 booklets that were out of date. With these statistics at hand, which plan would be the best choice for the company and why?

____

16. Rewrite the following formula to solve for degrees Rankine, which is the absolute temperature used with Fahrenheit:

$$°C = \frac{5}{9}(°R - 492).$$

____

17. Since the area (A) of a rectangle is found by multiplying its length (l) by its width (w), write a formula for finding the width of a rectangle when the area and length are known.

____

18. If a 2″ × 6″ joist has 8 board feet in it, how long is the piece?

____

19. What is the weight of a barrel of 40° API gravity oil?

____

20. How many 8-inch square tiles will be required to cover an entry that is 6 feet wide by 8 feet long?

____

6 PRACTICAL GEOMETRY

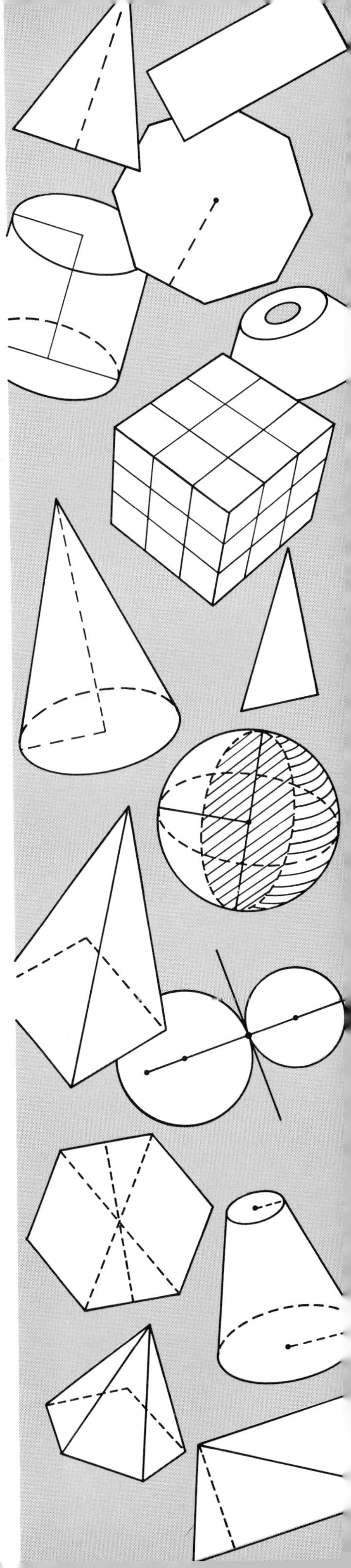

Objectives

Upon completion of chapter six, the student will be able to—

1. Find the perimeters of rectangles and other parallelograms, trapezoids, triangles, and polygons.
2. Find the circumferences of circles and ellipses.
3. Find the length of a triangle's side with the dimensions of two sides given.
4. Find the areas of parallelograms, trapezoids, triangles, regular polygons, circles, and ellipses.
5. Solve problems involving plane figures using geometrical formulas.
6. Find the volumes of rectangular solids, cylinders, elliptical solids, cones, pyramids, frustums of cones and pyramids, and spheres.
7. Find the lateral and total surface areas of some solid figures.
8. Solve problems involving solid figures using geometrical formulas.
9. Construct geometric figures using only a compass and a straightedge.
10. Use triangles and T-square properly for drawing geometric figures.

Geometry deals with the measurements, properties, and relationships of points, lines, angles, surfaces, and solids. *Plane* geometry is concerned with plane, or two-dimensional, figures. *Solid* geometry deals with solid, or three-dimensional, objects.

Using geometry to solve everyday problems calls for deductive reasoning—that is, reasoning or logic based on certain known facts called *theorems.* To solve a problem in geometry, the student must first have a clear definition of terms, must understand the theorems and their related formulas, and must be able to draw conclusions based on given facts. In pure geometry, the theorems are proved. This chapter will not concentrate on proving theorems. Rather, it will apply geometric formulas derived from the theorems to situations that might arise in shop and field.

An aid to visualizing a problem and its solution is a geometric construction; thus, the student should also learn to use the compass, straightedge, and basic drawing tools.

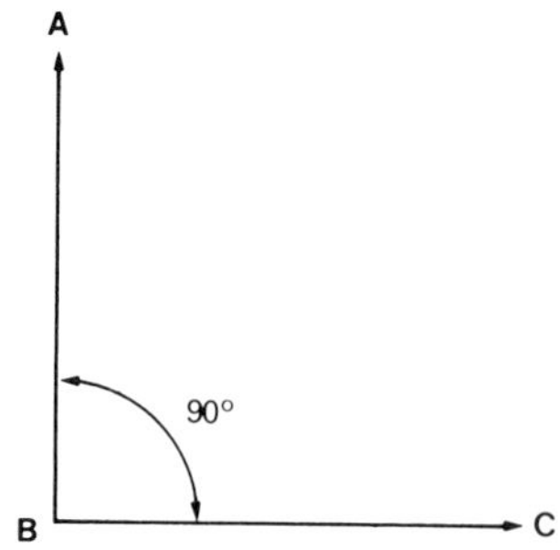

Fig. 6.1. A right angle

Plane Figures

A plane geometrical figure is a polygon or a circle. A *polygon* is any closed figure of three or more straight sides on the same plane. A polygon has length and width, but no depth. Measurements of length and width (or base and altitude, or height) and angles are used to determine the area and perimeter of a polygon. A *circle* is a closed plane curve, all points of which are equally distant from a point within called the center. Circle measurements are concerned with circumference, radius, diameter, and area.

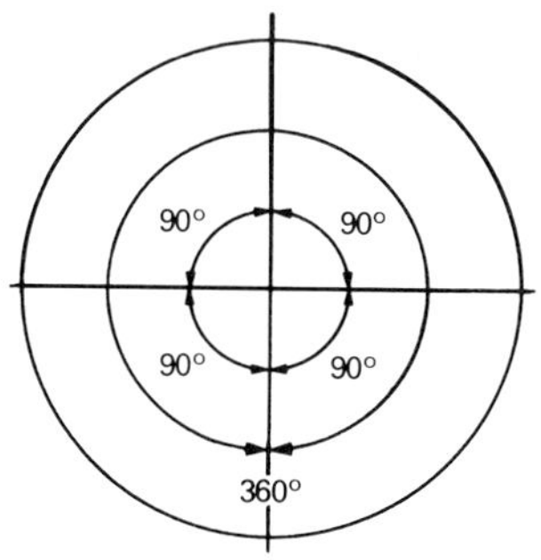

Fig. 6.2. A circle

ANGLES AND LINES

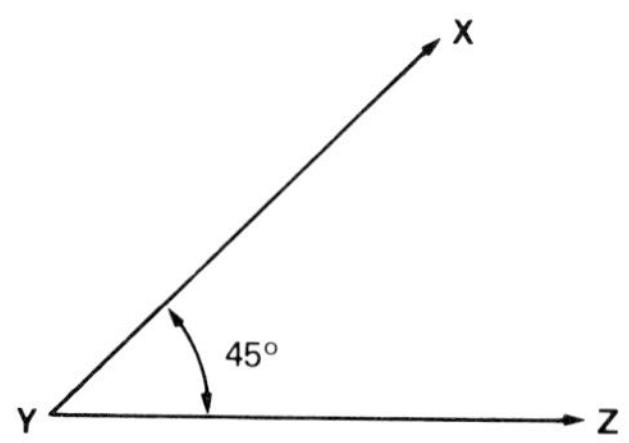

Fig. 6.3. Acute angle

An angle is formed when two straight lines, called the *sides,* meet at a point, called the *vertex*. For example, in figure 6.1 sides *AB* and *BC* meet at point *B,* forming the angle *ABC.* The size of an angle is measured in degrees (°). A circle contains 360°. The angle in figure 6.1 contains 90° and is called a *right angle.* If a circle is divided into four equal parts, it contains four right angles of 90° each, or 360° (fig. 6.2).

An angle that measures less than 90° is an *acute angle* (fig. 6.3), and an angle that measures more than 90° is an *obtuse angle* (fig. 6.4).

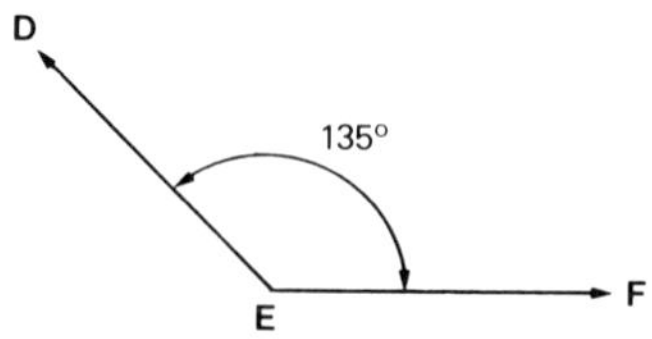

Fig. 6.4. Obtuse angle

Parallel lines are straight lines in the same plane that do not meet, or intersect, however far extended. They are equidistant from each other (fig. 6.5).

Lines that meet at a right angle are *perpendicular* to each other, as lines *AB* and *BC* in figure 6.1.

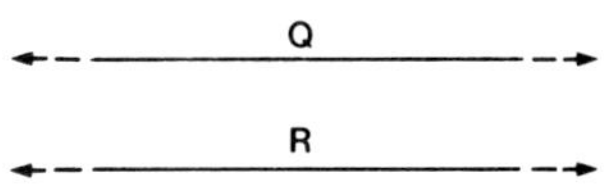

Fig. 6.5. Parallel lines

PARALLELOGRAMS

The parallelogram is a closed figure with opposite sides parallel (fig. 6.6). The sides of a parallelogram can meet at any angle. If the sides meet at right angles, that is, perpendicular to each other, then the parallelogram is a *rectangle* (fig. 6.7). If the sides are the same length and meet at right angles, the parallelogram is a *square* (fig. 6.8).

In the parallelogram in figure 6.6, line *CD* forms the base (length) and line *BD* is the height (width). In the rectangle, as in the square, line *CD* is the base and line *BD* is the height.

The perimeter of a plane figure is the distance around the figure, or the sum of the lengths of the sides. To find the **perimeter of the rectangle** shown in figure 6.7, add $AB + BD + DC + CA$:

$$8'' + 5'' + 8'' + 5'' = 28''.$$

Finding the perimeter of a square can be simplified by multiplying the length of one side times 4. For example, the perimeter of the square in figure 6.8 is equal to line $AB \times 4$:

$$5'' \times 4 = 20''.$$

The area of a figure is the amount of surface space enclosed within the boundary lines of that figure (fig. 6.9). To find the **area of a parallelogram,** multiply the number of units in its base times the number of units in its height to get the number of *square units* contained in the parallelogram ($A = bh$). To find the area of the rectangle in figure 6.9, multiply the base (13") times the height (4.5"):

$$A = bh$$
$$A = 13'' \times 4.5'' = 58.5 \text{ in.}^2$$

In computing areas, care must be taken to use the same units for base and height. For example, if the base is expressed in feet and the height in inches, the feet must be changed to inches or the inches to feet before multiplying. Table 6.1 and the tables in chapter four will be helpful with conversions used in area measurement.

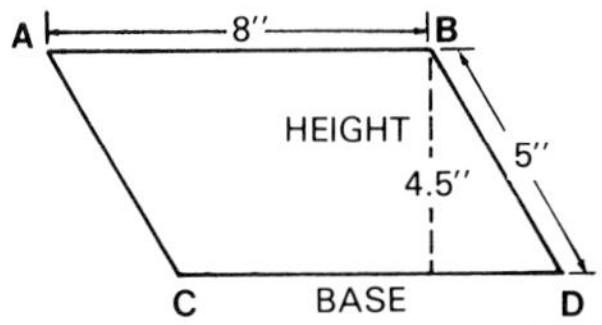

Fig. 6.6. Parallelogram

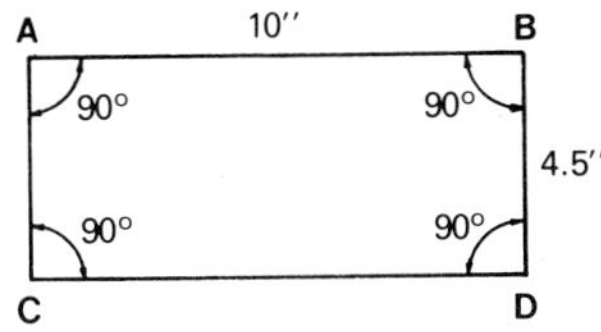

Fig. 6.7. Rectangle

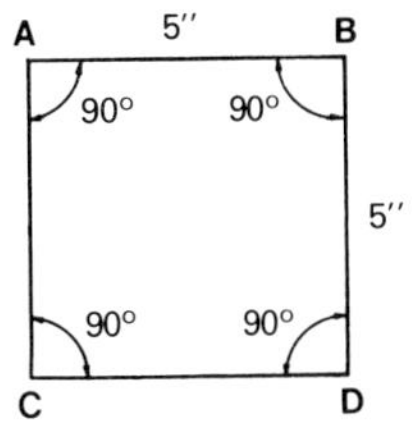

Fig. 6.8. Square

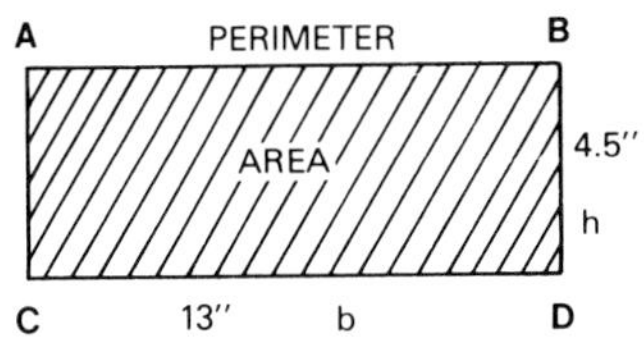

Fig. 6.9. Area and perimeter of a rectangle

TRAPEZOIDS

A trapezoid is a four-sided plane figure, two of whose sides are parallel (fig. 6.10). The **area of a trapezoid** is equal to one-half its height (h) times the sum of its bases (b and b'). To find the area of the trapezoid in figure 6.10, use the formula

$$A = \tfrac{1}{2}h \times (b + b').$$

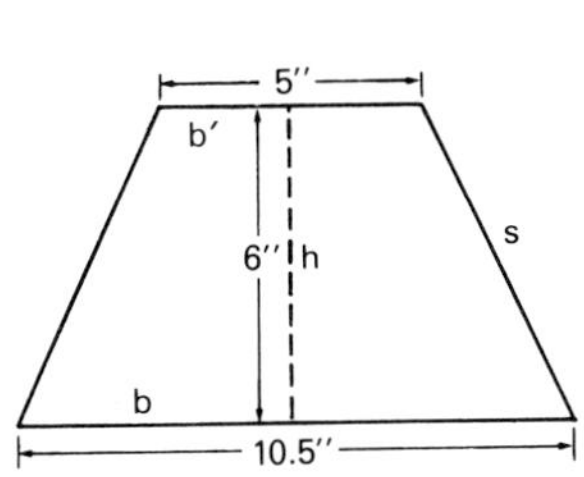

Fig. 6.10. Trapezoid

Substituting the measurements, the equation becomes

$$A = \tfrac{1}{2}(6) \times (10.5 + 5) = 3 \times 15.5 = 46.5 \text{ in.}^2$$

TRIANGLES

A triangle is a closed figure formed by three straight lines connecting three points not in a straight line. The points are called the *vertices* of the triangle. The lines between the vertices are the *sides.* In the triangle shown in figure 6.11, sides AB and BC are called *legs.* Side AC, which is the longest and is opposite the right angle, is called the *hypotenuse.* Because angle ABC has 90°, this triangle is a *right triangle.*

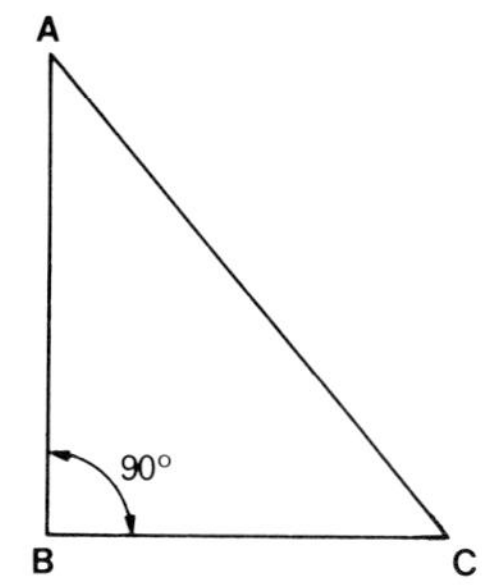

Fig. 6.11. Right triangle

An *equilateral* triangle is one with three equal sides as shown in figure 6.12 where AB equals BC equals CA. An *isosceles* triangle is one with two equal sides, or legs. Any one of the three sides of a triangle may be chosen as the base. In an isosceles triangle, however, *base* refers to the side that is not one of the legs. In figure 6.13, AB equals AC, and BC is the base. A *scalene* triangle is one with no two sides equal (fig. 6.14).

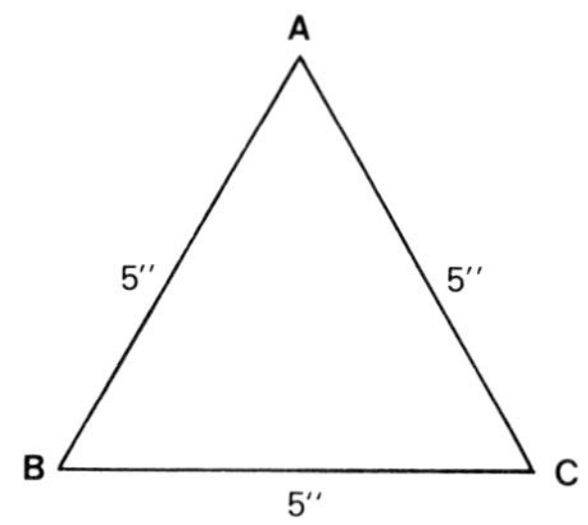

Fig. 6.12. Equilateral triangle

The **area of a triangle** is equal to one-half the product of its base and altitude. This statement, or theorem, can be written as a formula:

$$A = \tfrac{1}{2}bh$$

where

A = area
b = base
h = altitude, or height.

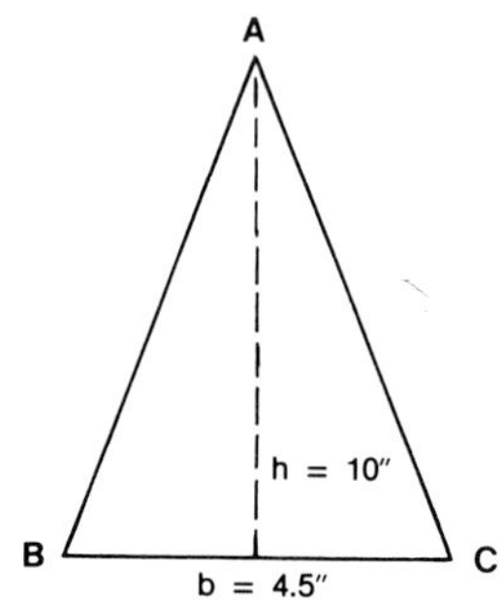

Fig. 6.13. Isosceles triangle

Using this formula, the area of the triangle in figure 6.13 can be found:

$$A = \tfrac{1}{2}(4.5'' \times 10'') = \tfrac{1}{2}(45) = 22.5 \text{ in.}^2$$

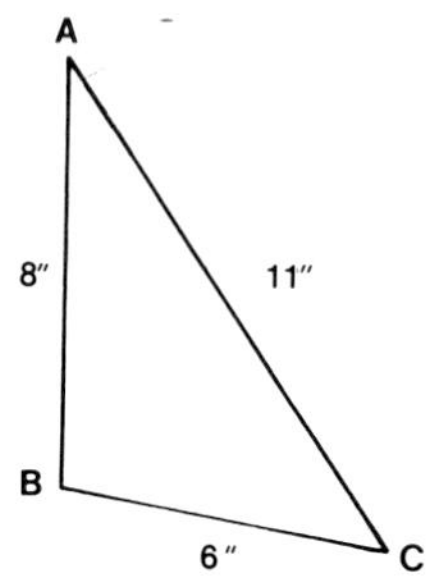

Fig. 6.14. Scalene triangle

This formula is logical since two identical triangles joined along equal sides always forms a parallelogram (fig. 6.15), and the area of any rectangle or parallelogram is equal to the product of its base and height ($A = bh$).

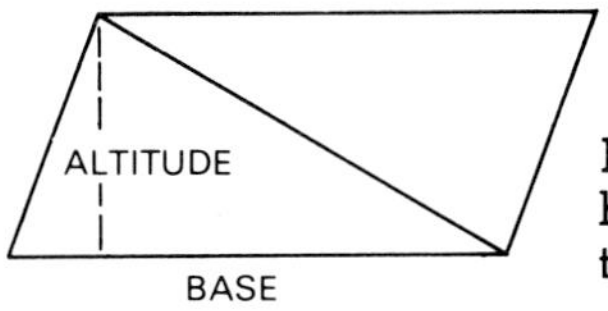

Fig. 6.15. Parallelogram has two identical triangles.

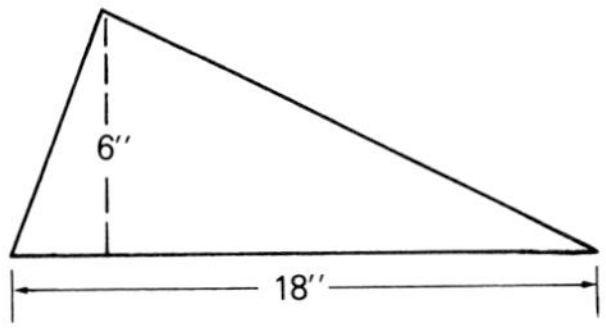

Fig. 6.16

Example Problem: Find the area of the triangle in figure 6.16.

Solution:

$A = \frac{1}{2}bh.$

$A = \frac{1}{2}(18'' \times 6'') = 108 \div 2 = 54 \text{ in.}^2$

As stated before, a right triangle contains one 90° angle. The area of a right triangle is calculated in the same manner as that of any triangle; however, the sides of a right triangle bear a peculiar relationship to one another, permitting the calculation of the length of one side if the other two sides are known. This relationship, or theorem, can be stated: *The square of the hypotenuse is equal to the sum of the squares of the other two sides (legs).* This relationship becomes apparent when the squares of a, b, and c are shown as areas (fig. 6.17).

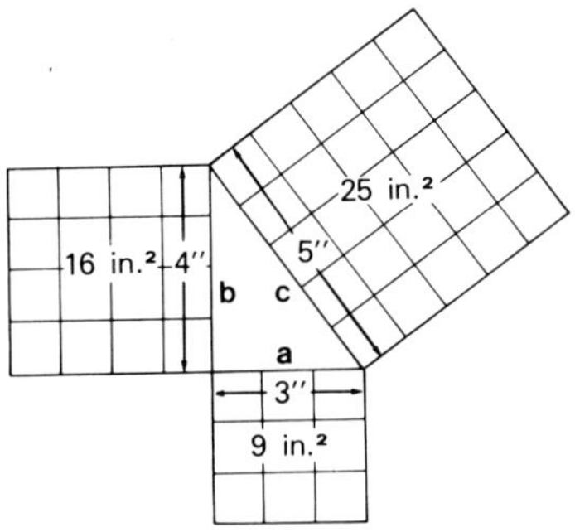

Fig. 6.17. Diagram showing that $c^2 = a^2 + b^2$

Example Problem: Find the hypotenuse of the triangle in figure 6.18.

Solution:

$c^2 = a^2 + b^2$

$c^2 = 3^2 + 4^2 = 9 + 16 = 25''$

$c = \sqrt{25} = 5''.$

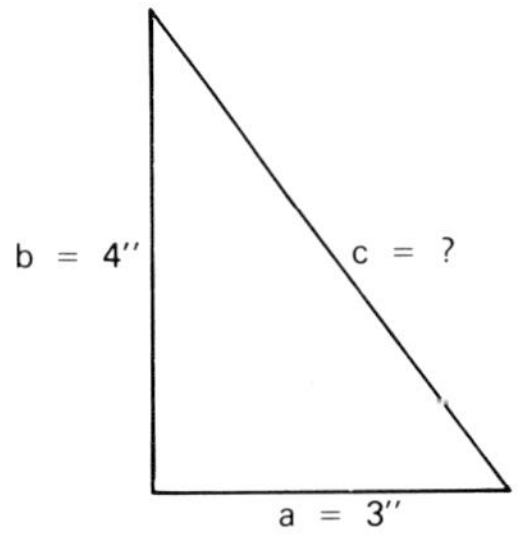

Fig. 6.18

To solve for an unknown leg, the equation is transposed to read

$$a^2 = c^2 - b^2$$

or

$$b^2 = c^2 - a^2.$$

Example Problem: Solve for leg b of the triangle in figure 6.19.

Solution:

$b^2 = c^2 - a^2$

$b^2 = 5^2 - 3^2 = 25 - 9 = 16''$

$b = \sqrt{16} = 4''.$

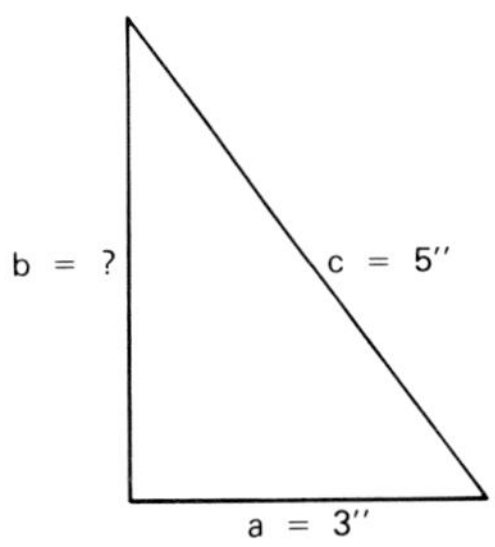

Fig. 6.19

OTHER POLYGONS

As stated earlier, a *regular polygon* is a closed figure with equal sides and equal angles. In addition to equilateral triangles (three-sided) and squares (four-sided), regular polygons can be pentagons (five-sided), hexagons (six-sided), heptagons (seven-sided), octagons (eight-sided), and so forth. Of course, **the perimeter of a regular polygon,** regardless of the number of sides, is determined by adding the measurements of each side or multiplying the number of sides times the length of one side. For example, the perimeter of the hexagon in figure 6.20 would be

$$P = 6 \times 4 \text{ mm} = 24 \text{ mm}.$$

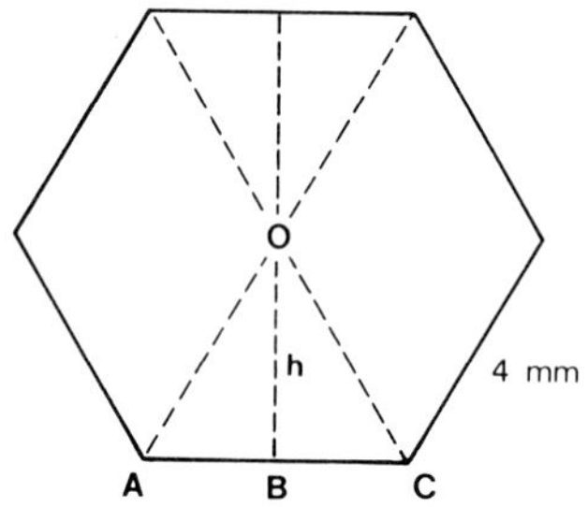

Fig. 6.20. Regular six-sided polygon (hexagon)

The **area of a polygon** is the number of square units contained in the surface bounded by its perimeter. One formula used to find the area of a regular polygon is

$$A = \tfrac{1}{2}hP$$

where

A = area
h = altitude of the polygon
P = perimeter of the polygon.

In figure 6.20, the altitude (h) is the perpendicular line OB, drawn from the center O to bisect side AC.

Example Problem: Find the area of the octagon shown in figure 6.21.

Solution: First, find the perimeter of the octagon by multiplying the length of one side times the number of sides:

$$P = 6.6'' \times 8 = 52.8''.$$

Then solve for the area:

$$A = \tfrac{1}{2}hP$$
$$A = \tfrac{1}{2}(10 \times 52.8) = 528 \div 2 = 264 \text{ in.}^2$$

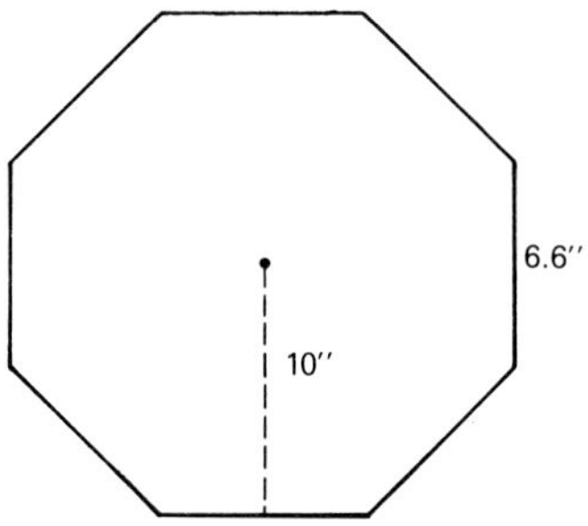

Fig. 6.21. Octagon

An *irregular polygon* has many sides of different lengths and different angles. To find the **perimeter of the polygon** shown in figure 6.22, add the lengths of the sides together:

$$3'' + 5'' + 3'' + 4'' + 5'' = 20'', \text{ or } 1'8''.$$

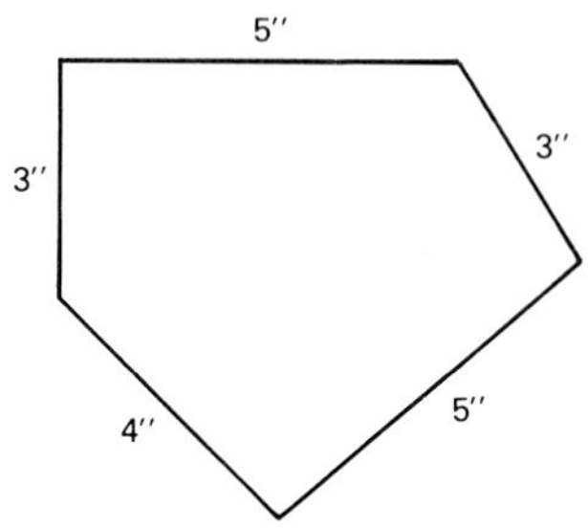

Fig. 6.22. Irregular five-sided polygon (pentagon)

The **area of an irregular polygon** can be found by constructing triangles within the polygon and using trigonometric functions (see chapter 7).

CIRCLES

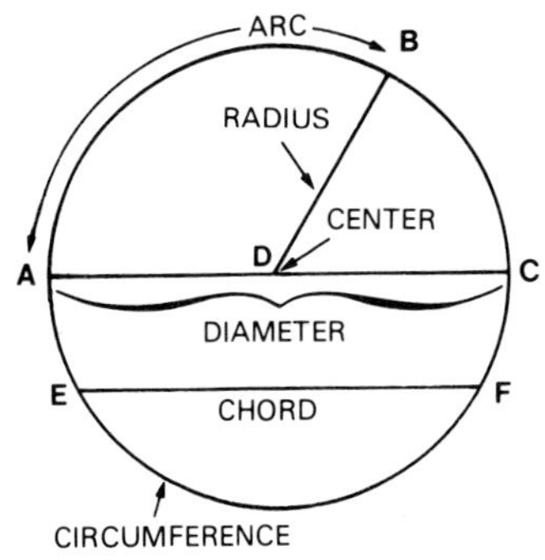

Fig. 6.23. Parts of a circle

A circle is a closed curve all points of which are in the same plane and are equally distant from a point within it called the *center.* The distance around the circle is called the *circumference.* A line from the center to any point on the circle is called a *radius.* In figure 6.23, lines *DB* and *DC* are radii, and all radii of the same circle are equal in length. A straight line drawn between any two points on the circle and passing through the center constitutes the *diameter* (line *AC*). A diameter is equal in length to two radii. All diameters of the same circle are equal. A straight line joining any two points on a circle is called a *chord (line EF*). Any part of the line forming the circle is called an *arc.* In figure 6.23, curves *AB, BC, CF,* and *AE* are all arcs.

Two or more circles having the same center but different radii are called *concentric circles* (fig. 6.24). A *semicircle* is half a circle (fig. 6.25).

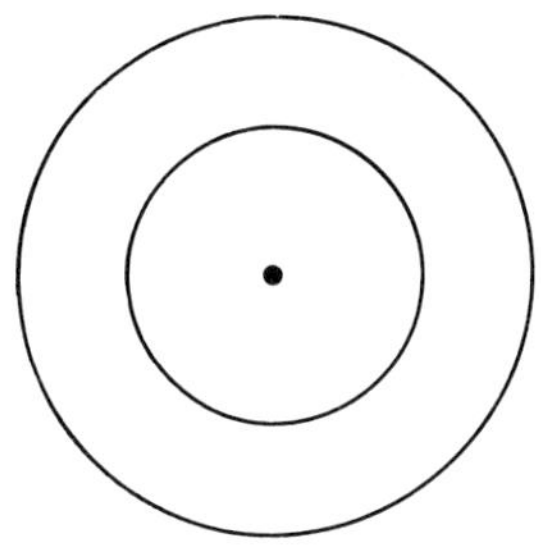
Fig. 6.24. Concentric circles

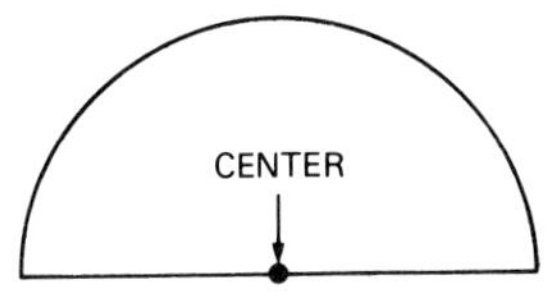

Fig. 6.25. Semicircle

The **circumference of a circle** is equal to its diameter (d) multiplied by the constant 3.141592654, generally shortened to 3.1416 and designated by the Greek letter π (*pi*). The equation for finding the circumference (C) is

$$C = \pi d, \text{ or } C = \pi 2r.$$

Example Problem: The radius of a circle is 8″. What is its circumference?

Solution:

$C = \pi 2r$
$C = 3.1416 \times 2 \times 8'' = 50.2656''$, or 50.27″.

The formula $C = \pi d$ can also be used to find a missing diameter or radius by transposing it to read

$$d = C/\pi.$$

Example Problem: The circumference of a circle is 15.708″. What is its diameter?

Solution:

$d = C/\pi$
$d = 15.708 \div 3.1416 = 5''$.

The **area of a circle** can be found by multiplying the square of its radius (r) by 3.1416 (π); that is,

$$A = \pi r^2.$$

Example Problem: What is the area of a circle whose radius is 6″?

Solution:

$A = \pi r^2$
$A = 3.1416 \times 6 \times 6 = 113.1 \text{ in.}^2$

If the area of a circle is known, then the formula $A = \pi r^2$ can be transposed to solve for the radius or diameter of a circle:

$$r^2 = A/\pi.$$
$$r = \sqrt{A/\pi}.$$
$$d = 2r.$$

Example Problem: Find the radius of a circle whose area is 254.4696 square inches.

Solution:

$r = \sqrt{A/\pi}.$
$r = \sqrt{254.4696 \div 3.1416} = \sqrt{81} = 9''.$

Example Problem: Find the diameter of the circle shown in figure 6.26.

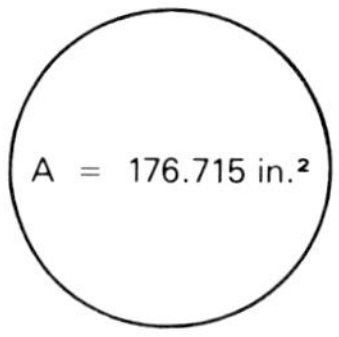

Fig. 6.26

Solution:

$r = \sqrt{A/\pi}.$
$r = \sqrt{176.715 \div 3.1416} = \sqrt{56.25} = 7.5''.$
$d = 2 \times 7.5 = 15''.$

ELLIPSES

An ellipse might be called a compressed circle. It is an oval-shaped closed curve with two axes, as shown in figure 6.27. The longer axis is called the *major* axis, and the shorter one the *minor* axis.

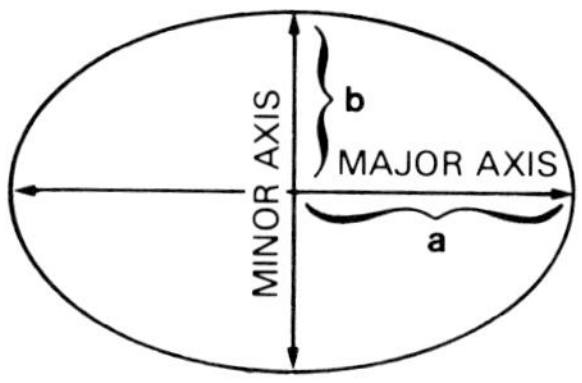

Fig. 6.27. Ellipse

Letting *a* represent half of the major axis and *b* half of the minor axis, the formula for finding the **perimeter of an ellipse** is

$$P = 2\pi\sqrt{(a^2 + b^2) \div 2.}$$

Example Problem: Find the perimeter of an ellipse whose major axis is 10 inches and whose minor axis is 5 inches.

Solution:

$a = \frac{1}{2} \text{ of } 10 = 5''; \ b = \frac{1}{2} \text{ of } 5 = 2.5''.$
$P = 2\pi\sqrt{(a^2 + b^2) \div 2.}$
$P = 2 \times 3.1416 \times \sqrt{(5^2 + 2.5^2) \div 2}$
$= 6.2832 \times \sqrt{(25 + 6.25) \div 2}$
$= 6.2832 \times \sqrt{15.625}$
$= 6.2832 \times 3.953 = 24.836''.$

The **area of an ellipse** is equal to one-half the major axis times one-half the minor axis times pi. With a and b representing half the major axis and minor axis respectively, the equation for finding the area of an ellipse is

$$A = \pi ab.$$

Note that this formula resembles the formula for finding the area of a circle; however, the circle has one radius (r) while the ellipse has two dimensions (a and b) to deal with.

Example Problem: What is the area of the ellipse shown in figure 6.28?

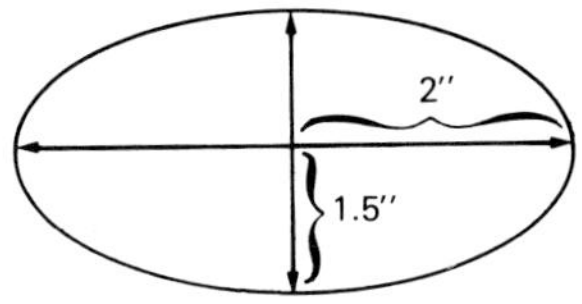

Fig. 6.28

Solution:

$A = \pi ab$
$A = 3.1416 \times 2 \times 1.5 = 9.4248 \text{ in.}^2$

In problems such as this, any two of the numbers may be multiplied together first, then the product of those two numbers is multiplied by the remaining number. For example,

$3.1416 \times 2 = 6.2832$, and $6.2832 \times 1.5 = 9.4248$

Likewise,

$2 \times 1.5 = 3$, and $3 \times 3.1416 = 9.4248$.

Table 6.1
Common Units of Area Measurement

1 ft^2	=	144 $in.^2$
1 yd^2	=	9 ft^2
1 rod^2	=	30.25 yd^2
1 acre	=	160 $rods^2$
1 acre	=	4,840 yd^2
1 $mile^2$	=	1 section
1 section	=	640 acres
1 hectare	=	2.471 acres

Table 6.2
Formulas for Finding Areas of Plane Figures

Figure	Formula	Meaning of Terms
Parallelogram, rectangle, square	$A = bh$	b = base h = height
Trapezoid	$A = \frac{1}{2}h(b + b')$	h = height b = bottom base b' = top base
Triangle	$A = \frac{1}{2}bh$	b = base h = altitude, or height
Other regular polygon (hexagon, octagon, etc.)	$A = \frac{1}{2}aP$	a = altitude P = perimeter
Circle	$A = \pi r^2$	π = 3.1416 r = radius
Ellipse	$A = \pi ab$	π = 3.1416 a = one-half the major axis b = one-half the minor axis

PRACTICE PROBLEMS

Refer to tables 6.1 and 6.2 for area measurements and formulas. If conversion factors given in Appendix C-1 are used, answers may vary.

1. Find the number of square feet in the trapezoid below. ____________

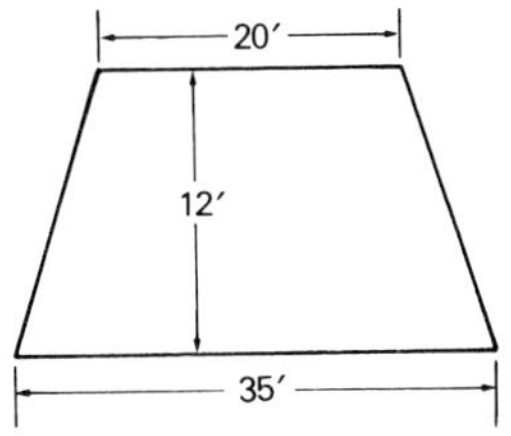

2. What is the area of the parallelogram below, in square feet? ____________

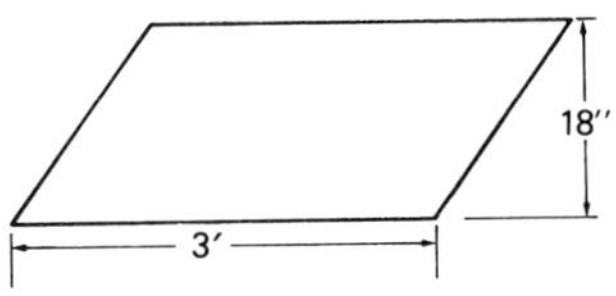

3. What is the perimeter of a rectangle that measures 7 inches by 14 inches? ____________

4. Find the number of acres in a field that is 2,700 feet long and 600 feet wide. (1 acre = 43,560 square feet.) ____________

5. How many square yards are there in a rectangle that is 125 feet long and 80 feet wide? ____________

   What is the perimeter in yards? ____________

6. A sheet of packing weighs 1.25 pounds per square foot. How much will a piece of this packing that is 8 feet long and 30 inches wide weigh? ____________

7. A saltwater pit is 250 feet long and 130 feet wide. How many acres does it cover? Round off the answer to the nearest common fraction. ____________

8. To reach a well location in a rice field, it was necessary to build a board road 1,400 feet long and 8 feet wide.

   a. How many square feet of lumber were required? ____________

   b. If the boards are 1 foot wide, how many 8-foot boards will be needed? ____________

9. You are going to build a new bed for mounting on a pickup truck. The bed is to be 8 feet long × 54 inches wide × 22¼ inches deep. Without allowing for waste, how many square feet of steel will be needed to make the floor, two sides, and tailgate? ____________

10. Find the areas of the triangles below.

a. ____________

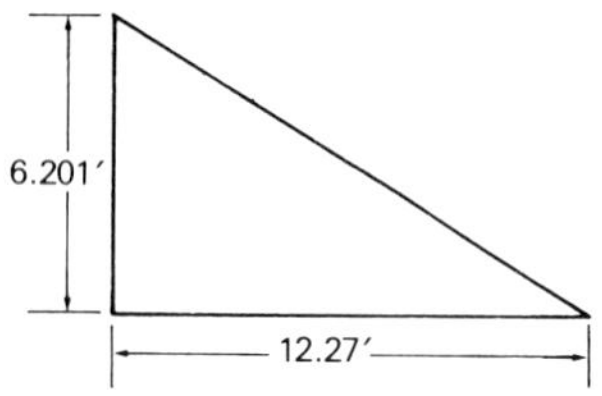

b. ____________

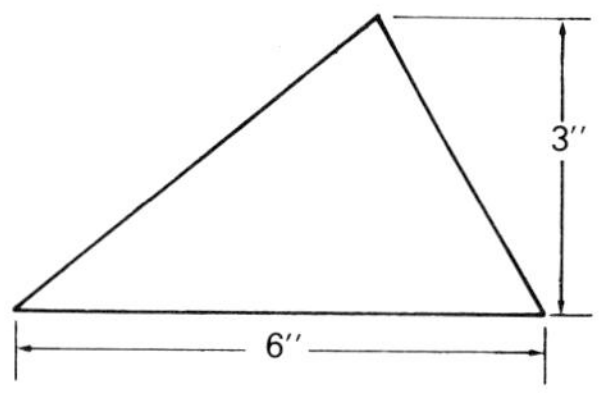

c. $b =$ ____________ ____________

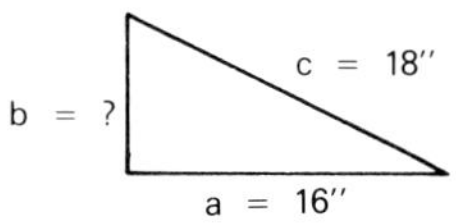

11. Find the area of the shaded portion of the circle below. ____________

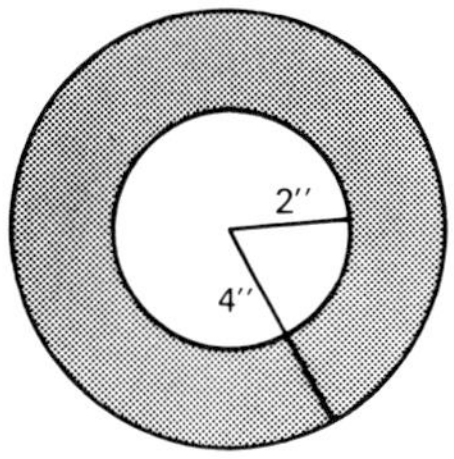

12. The tank shown at right has a 7-foot outside diameter and is banded with eight bands that overlap 6 inches. How many feet of band iron were required to make these bands? ____________

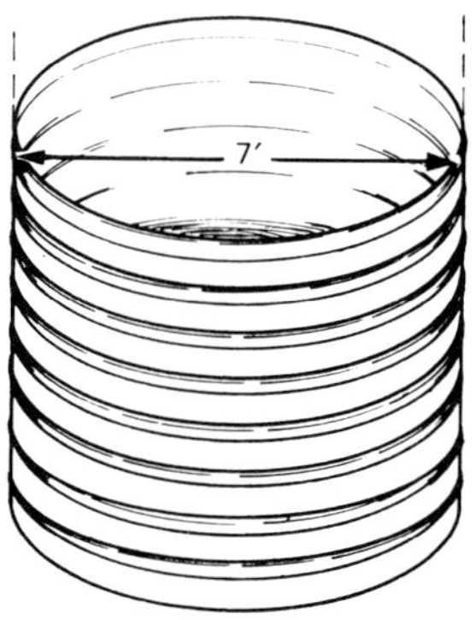

13. In landscaping a camp, an elliptical plot for a swimming pool is to be laid out. The major axis of the ellipse will be 80 feet, and the minor axis will be 60 feet. What part of an acre will the plot contain? (1 acre = 43,560 square feet.) ____________

14. Calculate the total area of the irregular shape below. ____________

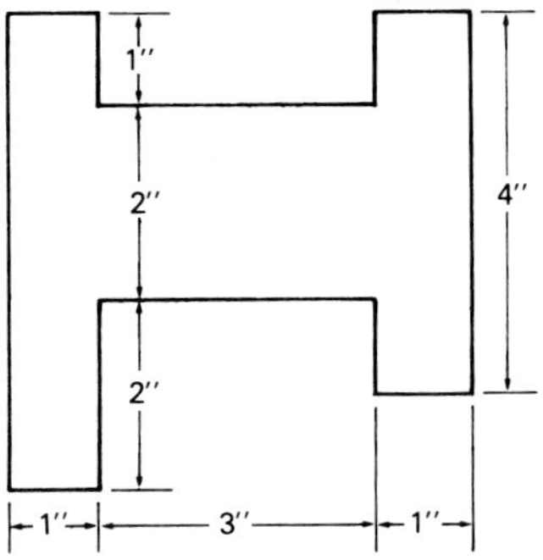

15. The tube sheet shown in the sketch below is 36 inches in diameter and is drilled for thirty 3-inch holes.

    a. How many square inches of metal were removed? ____________

    b. How many square inches of metal are left in the sheet? ____________

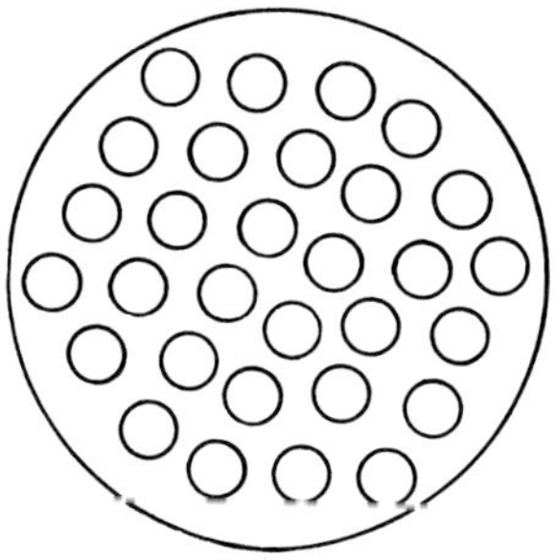

Solid Figures

Plane surfaces have two dimensions: length and width. In solid figures, another dimension—thickness—must be taken into account. The amount of space a solid figure occupies is called its *volume.* Volumes, which are obtained by dealing with all three dimensions, are expressed in cubic inches, cubic feet, cubic yards, cubic metres, and so forth. Just as in calculating dimensions of plane figures, all three dimensions in the calculation of volumes must be expressed in the same units. Inches cannot be multiplied by feet, feet by yards, or gallons by barrels.

RECTANGULAR SOLIDS

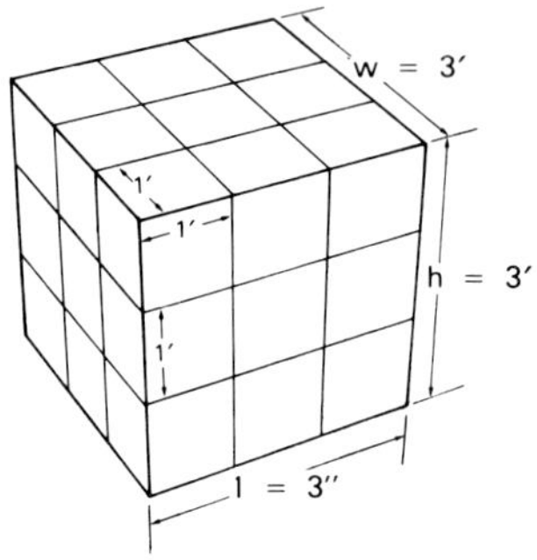

Fig. 6.29. Cube

The rectangular solid shown in figure 6.29, called a cube when all sides are equal, illustrates cubic measure. Multiplying the three dimensions of the solid—*length* times *width* times *height*—gives its *volume:*

$$V = lwh.$$
$$V = 3' \times 3' \times 3' = 27 \text{ ft.}^3$$

By marking the block with lines 1 foot apart as shown and sawing along these lines, it is evident that the block would be divided into twenty-seven smaller blocks, each 1 foot high, 1 foot wide, and 1 foot deep.

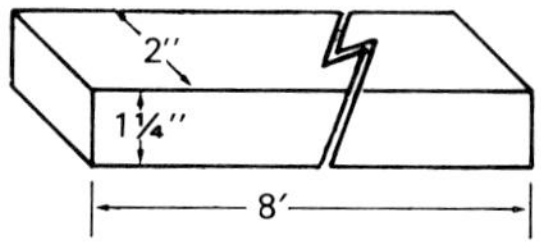

Fig. 6.30. Rectangular solid

Example Problem: What is the volume of the rectangular solid shown in figure 6.30? Express the answer in cubic inches.

Solution:

$l = 8' = 96''$; $w = 2$; $h = 1\frac{1}{4}''$.

$V = lwh$

$V = 96'' \times 2'' \times 1\frac{1}{4}'' = 240 \text{ in.}^3$

Volumes are frequently expressed in gallons or barrels, and to solve for them, measurement conversions given in table 6.3 are necessary. These conversions change cubic inches and cubic feet into gallons and barrels.

Table 6.3
Common Units of Volume Measurement

1 ft^3	=	1,728 $in.^3$
1 m^3	=	35.31 ft^3
1 yd^3	=	27 ft^3
1 ft^3	=	7.48 gal
1 m^3	=	264.12 gal
1 bbl	=	42 gal
1 m^3	=	6.29 bbl
1 ft^3	=	0.178 bbl
1 bbl	=	5.61 ft^3

Example Problem: How many barrels of water will it take to fill a rectangular tank that is 7 feet long, 4 feet wide, and 3½ feet high?

Solution:

$V = lwh$

$V = 7' \times 4' \times 3\frac{1}{2}' = 98 \text{ ft.}^3$

If 5.61 cubic feet equals 1 barrel, then 98 cubic feet equals 17.47 barrels:

$98 \div 5.61 = 17.47$ barrels.

To find the **total surface area,** find the area of each side and add the areas together. The answer will be in square measurements. For example, to find the total surface area of the rectangular solid in figure 6.30, find the area of the top and multiply by 2 to find the area of the top and the bottom:

$$A = lw$$
$$A = 96'' \times 2'' = 192 \times 2 \text{ sides} = 384 \text{ in.}^2$$

Then find the area of the two sides:

$$A = 96'' \times 1\tfrac{1}{4}'' = 120 \times 2 \text{ sides} = 240 \text{ in.}^2$$

And the area of the two ends:

$$A = 2'' \times 1\tfrac{1}{4}'' = 2\tfrac{1}{2} \times 2 \text{ sides} = 5 \text{ in.}^2$$

Finally, add the three dimensions to obtain the total surface area (T):

$$T = 384 + 240 + 5 = 629 \text{ in.}^2$$

CYLINDERS

In industry, it is frequently necessary to find the volume of cylinders such as storage tanks, pipes, and pump cylinders. These volumes are usually expressed in gallons or barrels and require the appropriate measurement conversions.

The **volume of a cylinder** (fig. 6.31) is equal to the product of the area of its base (B) times the height (h):

$$V = B \times h.$$

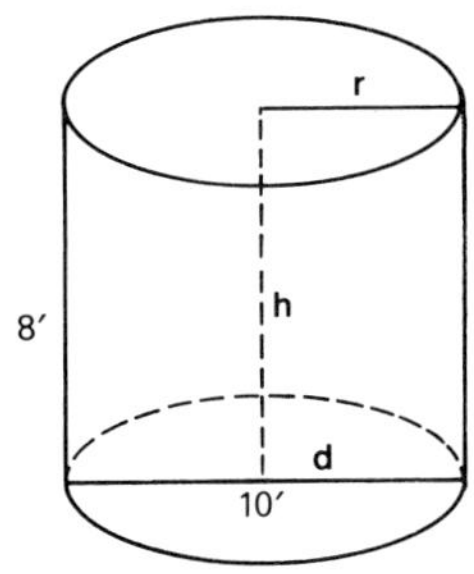

Fig. 6.31. Cylinder

Remembering that the area of a circle is equal to πr^2, the area of the cylinder base becomes πr^2, and the formula for finding the volume of a cylinder becomes

$$V = \pi r^2 h.$$

Example Problem: Find the volume of the cylinder shown in figure 6.31. Its diameter is 10 feet, and its height is 8 feet.

Solution:

$r = \tfrac{1}{2}$ of $10' = 5'$; $h = 8'$.
$V = \pi r^2 h$
$V = 3.1416 \times 5' \times 5' \times 8' = 628.32 \text{ ft}^3.$

Example Problem: How many barrels of oil can be stored in a tank that is 20 feet in diameter and 10 feet high?

Solution:

$r = \tfrac{1}{2}$ of $20' = 10'$; $h = 10'$.
$V = \pi r^2 h$
$V = 3.1416 \times 10' \times 10' \times 10' = 3{,}141.6 \text{ ft}^3.$
If 1 cubic foot equals 7.48 gallons, then 3,141.6 cubic feet equals 23,499 gallons; and if 42 gallons equals 1 barrel, then

$23{,}499 \div 42 = 559.5$ bbl.

ELLIPTICAL SOLIDS

An elliptical solid is like a cylinder except that the base is an ellipse instead of a circle. The equation for finding the area of an ellipse ($A = \pi ab$) is used to find the volume of an elliptical solid. Referring to figure 6.32, the formula for finding the volume of an elliptical solid can be written as

$$V = \pi abh$$

where

a = one-half the major axis
b = one-half the minor axis
h = height.

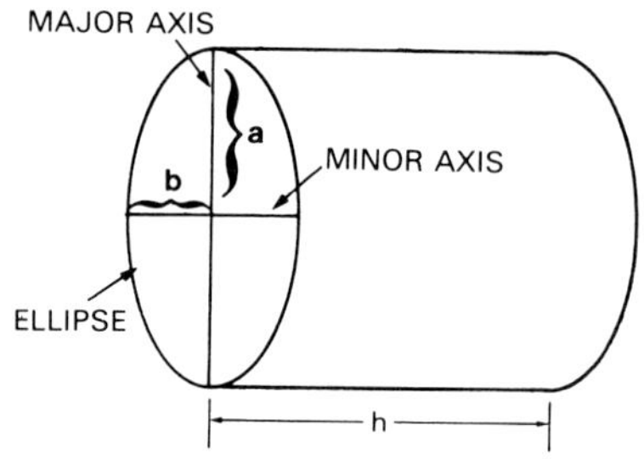

Fig. 6.32. Elliptical solid

Example Problem: Find the volume of an elliptical tank whose major axis is 8 feet, whose minor axis is 6 feet, and whose height is 10 feet.

Solution:

$a = 4'; b = 3'; h = 10'.$
$V = \pi abh$
$V = 3.1416 \times 4 \times 3 \times 10 = 376.992$, or 377 ft^3.

Example Problem: A tank truck has an elliptical tank that is 14 feet long. The major and minor axes of the elliptical ends of the tank are 8 feet and 6 feet. How many gallons of gasoline will the tank hold?

Solution:

$a = 4'; b = 3'; h = 14'.$
$V = \pi abh$
$V = 3.1416 \times 4 \times 3 \times 14 = 527.789$ ft^3.

If 1 cubic foot equals 7.48 gallons, then

$527.79 \times 7.48 = 3{,}947.86$ gal.

CONES

The most common type of cone—the right circular cone—is a solid generated by rotating a right triangle about one of its legs (fig. 6.33). The circle at the bottom of the cone is its base, and its altitude, or height, is the distance between the center of the base and the apex.

The **volume of a cone** is equal to one-third of the area of the base times the altitude:

$$V = \tfrac{1}{3}(Bh)$$

or

$$V = \tfrac{1}{3}(\pi r^2 h).$$

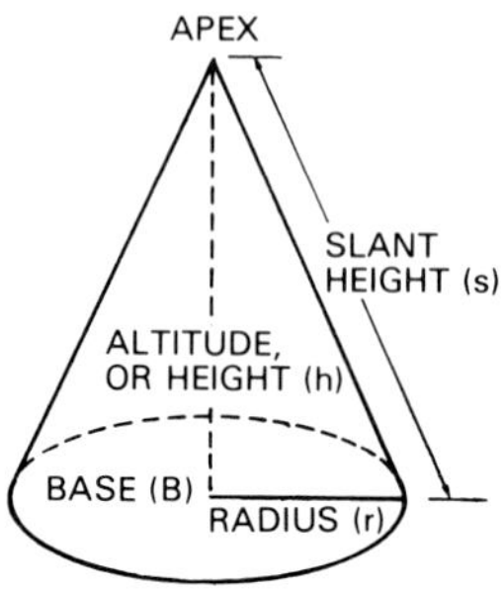

Fig. 6.33. Cone

Example Problem: Find the volume of a cone whose base has a 6-inch diameter and whose altitude is 7 inches.

Solution:

$V = \frac{1}{3}(\pi r^2 h)$

$V = \frac{1}{3}(3.1416 \times 3^2 \times 7)$

$= 197.9208 \div 3 = 65.9736 \text{ in.}^3$

Probably more important to industry than the volume of a cone is its lateral area, or conical surface (fig. 6.34). The **lateral area of a right circular cone** equals half the product of the slant height and the circumference of the base:

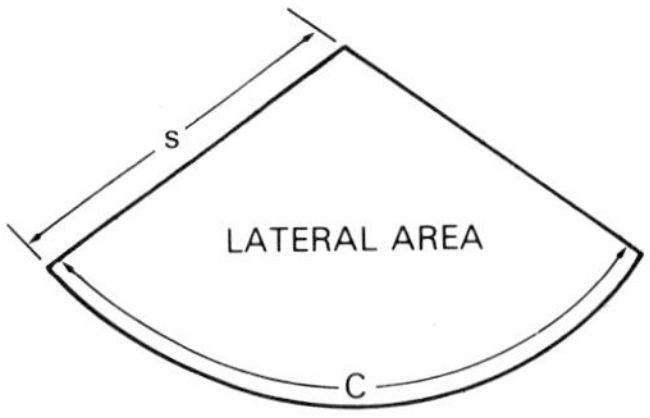

Fig. 6.34. Lateral area of a cone, or conical surface

$$L = \frac{1}{2}sC$$

where

L = lateral area
s = slant height
C = circumference of the base.

Since $C = 2\pi r$, then

$$L = \frac{s \times \not{2}\pi r}{\not{2}} = \pi rs.$$

Example Problem: Find the lateral area of a cone whose slant height is 5 millimetres and whose base radius is 2 millimetres.

Solution:

$L = \pi rs$

$L = 3.1416 \times 2 \times 5 = 31.416 \text{ mm}^2.$

If the altitude and base radius of a cone are known, the slant height can be calculated by using the formula for finding the hypotenuse of a triangle:

$$a^2 + b^2 = c^2.$$

Example Problem: What is the lateral area of the cone depicted in figure 6.35?

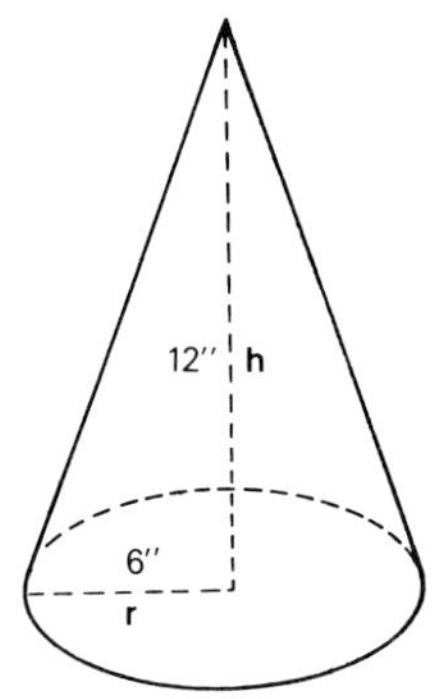

Fig. 6.35

Solution: First, find the slant height (s) of the cone, substituting h for a and r for b in the equation above:

$s^2 = h^2 + r^2$

$s^2 = 12^2 + 6^2 = 144 + 36 = 180$

$s = \sqrt{180} = 13.4164''.$

Then solve for the lateral area:

$L = \pi rs$
$L = 3.416 \times 6 \times 13.4164 = 252.894\ \text{in.}^2$

PYRAMIDS

A pyramid is a solid figure that has a polygon for its base and triangles with a common vertex for its faces (fig. 6.36). A regular pyramid may have a square for a base, or it may have any regular polygon such as a pentagon, hexagon, or octagon. Remember, a regular polygon has equal sides and equal angles.

The **volume of a pyramid** equals one-third the product of its base area (B) and its altitude (h):

$$V = \tfrac{1}{3}(Bh).$$

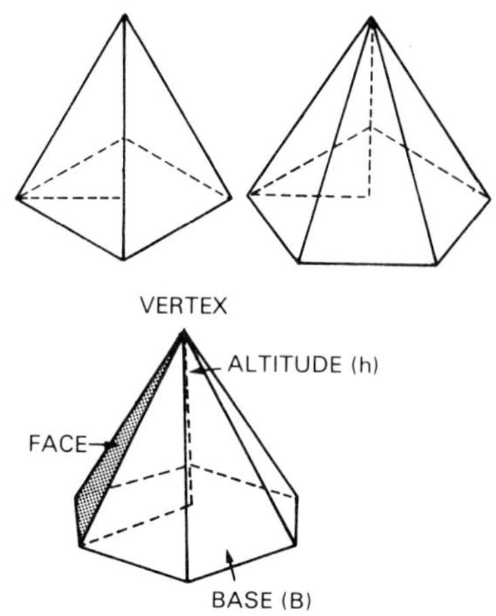

Fig. 6.36. Pyramids

Example Problem: Find the volume of the pyramid shown in figure 6.37.

Solution:

$B = 2' \times 2' = 4\ \text{ft}^2;\ h = 3'.$
$V = \tfrac{1}{3}(Bh).$
$V = (4 \times 3) \div 3 = 4\ \text{ft}^3.$

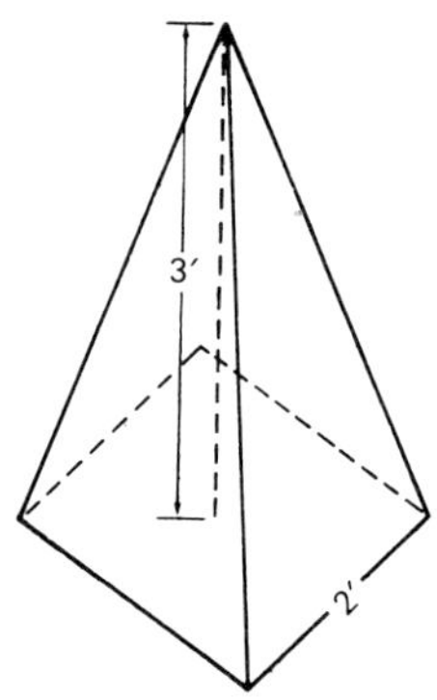

Fig. 6.37

To find the **lateral area of a pyramid,** the areas of the triangles that make up the lateral surface can be found and added together, or a formula can be used involving the pyramid's slant height and perimeter. In a regular pyramid (one whose base is a regular polygon), the altitudes of these triangles from their common vertex to their bases are the same as the pyramid's slant height (fig. 6.38). The sum of the sides of the polygon that form the pyramid base equals its perimeter. Thus, using the slant height and the perimeter, the lateral area of a pyramid can be found using the formula

$$L = \tfrac{1}{2}(sP)$$

where

L = lateral area
s = slant height
P = perimeter of the base.

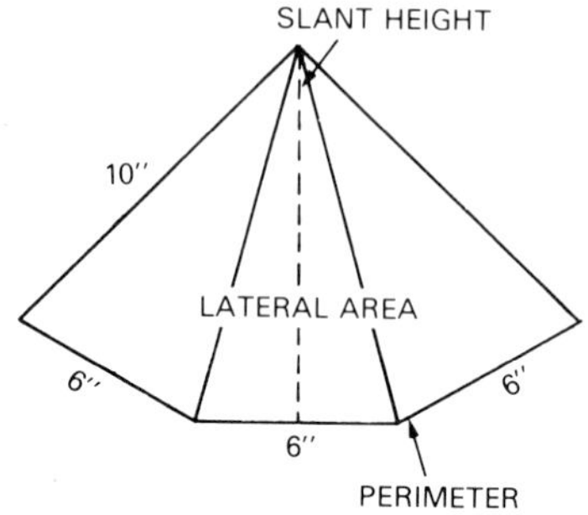

Fig. 6.38

Example Problem: Find the lateral area of a pyramid whose slant height is 10 inches and whose base is a hexagon with 6-inch sides.

Solution:

$L = \tfrac{1}{2}(sP)$
$L = 10'' \times 6'' \times 6\ \text{sides} \div 2 = 180\ \text{in.}^2$

The **total area of a pyramid,** like the **total area of a cone,** is found by adding the lateral area and the base area. For regular pyramids and right circular cones, the total area can be calculated with the formula

$$T = \pi r^2 + \pi rs$$

or

$$T = \pi r(r + s)$$

where

T = total area
r = radius of the base
s = slant height.

This formula is a combination of the formulas for calculating the lateral area and the base area.

Example Problem: Find the total area of a cone with a base radius of 1.5 feet and a slant height of 30 inches.

Solution: Before calculations can begin, the measurements must be converted to inches or feet. Using feet as the measurement,

$30'' \div 12'' = 2.5'$.

Then

$T = \pi r(r + s)$
$T = 3.1416 \times 1.5 \times (1.5 + 2.5) = 18.8496 = 18.85\ ft^2$.

FRUSTUMS

When a pyramid or a cone is cut off at any point below the apex by a plane that is parallel to the base, the portion below the cutting plane is called a frustum of a cone (fig. 6.39) or a frustum of a pyramid (fig. 6.40). These shapes are commonly seen in concrete bases or in the welded portion of a pipe swedge.

The base of a pyramid or cone (B) and the section of the intersecting plane (B') are called the *bases* of the frustum; the rest of the surface is called the *lateral surface* (L). The altitude (h) of a frustum is the distance between the centers of the two bases. This altitude line is perpendicular to both bases.

The formula for finding the **lateral area of a frustum of a cone** is half the slant height times the sum of the circumferences of the bases:

$$L = \tfrac{1}{2}s(C + C')$$

where

L = lateral area
s = slant height of the frustum
C = circumference of the cone base
C' = circumference of the section base.

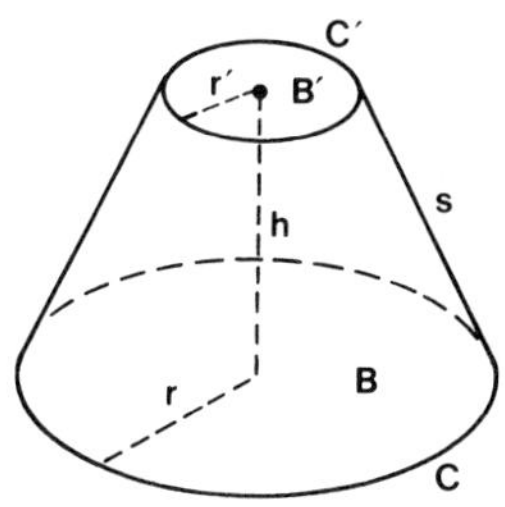

Fig. 6.39. Frustum of a cone

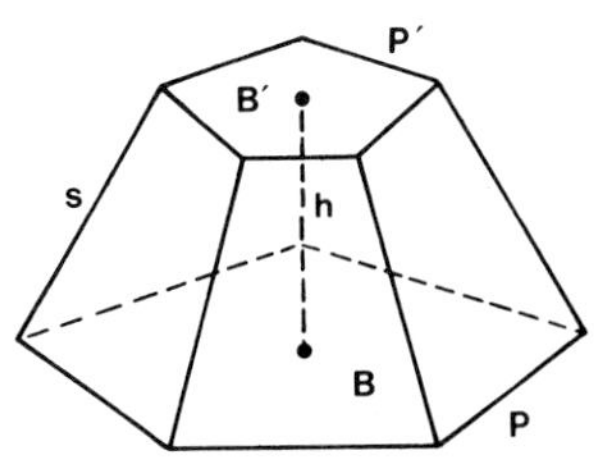

Fig. 6.40. Frustum of a pyramid

Example Problem: Find the lateral area of the frustum shown in figure 6.41.

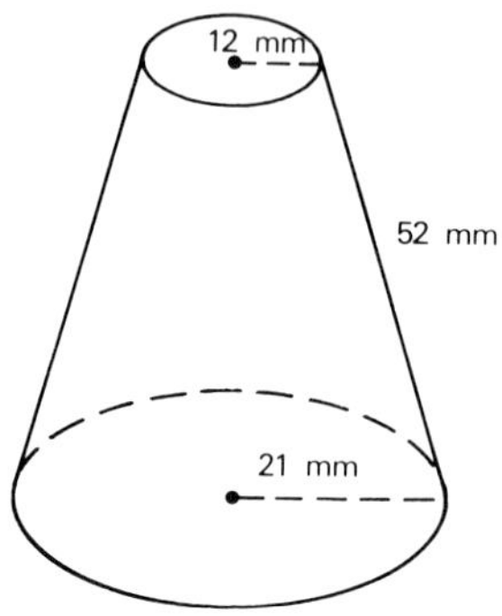

Fig. 6.41

Solution: Using the radii given, find the circumferences C and C':

$$C = \pi 2r$$
$$C = 3.1416 \times 2 \times 21 \text{ mm} = 131.95 \text{ mm}.$$
$$C' = 3.1416 \times 2 \times 12 \text{ mm} = 75.40 \text{ mm}.$$

Then solve for the lateral area:

$$L = \tfrac{1}{2}s(C + C')$$
$$L = 52 \times (131.95 + 75.40) \div 2 = 5{,}391.1 \text{ mm}^2.$$

The formula for finding the **lateral area of the frustum of a pyramid** is the same as that for a cone, except perimeters are used instead of circumferences:

$$L = \tfrac{1}{2}s(P + P')$$

where

L = lateral area
s = slant height of the frustum
P = perimeter of the pyramid base
P' = perimeter of the section base.

The **volume of a frustum of a cone or a pyramid** is obtained by the formula

$$V = \tfrac{1}{3}h(B + B' + \sqrt{BB'})$$

where

V = volume
h = height of the frustum
B = area of the cone base
B' = area of the section base.

This formula can be used whether the frustum is regular or not; that is, the bases do not have to be perfect circles or even-sided polygons.

Example Problem: Find the volume of the frustum shown in figure 6.42. Give your answer in cubic feet.

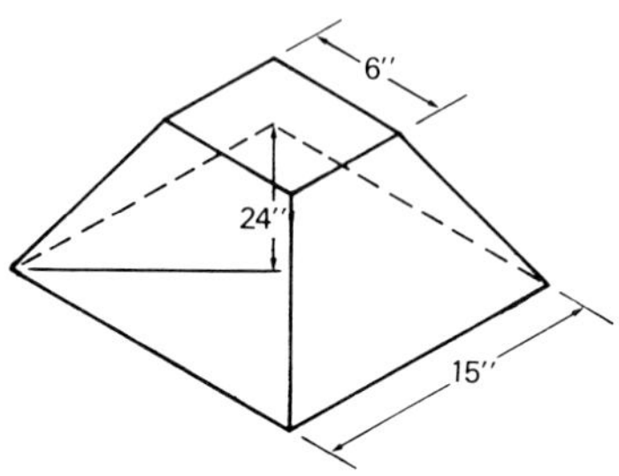

Fig. 6.42

Solution: First, solve for base areas:

$$B = 15'' \times 15'' = 225 \text{ in.}^2$$
$$B' = 6'' \times 6'' = 36 \text{ in.}^2$$

Then use the formula to find volume:

$$V = \tfrac{1}{3}h(B + B' + \sqrt{BB'})$$
$$V = 24 \div 3 \times (225 + 36 + \sqrt{225 \times 36})$$
$$= 8 \times (261 + 90) = 8 \times 351 = 2{,}808 \text{ in.}^3$$

Since 1 cubic foot equals 1,728 cubic inches, then
$2{,}808 \div 1{,}728 = 1.625 \text{ ft}^3$.

SPHERES

A sphere is a closed solid whose points are all equally distant from a point within called its center (fig. 6.43). A sphere is different from a circle in that a sphere exists in space while a circle is flat. It is helpful to think of a sphere as the shape of a basketball or a globe. Both the sphere and the circle have centers, diameters, and radii.

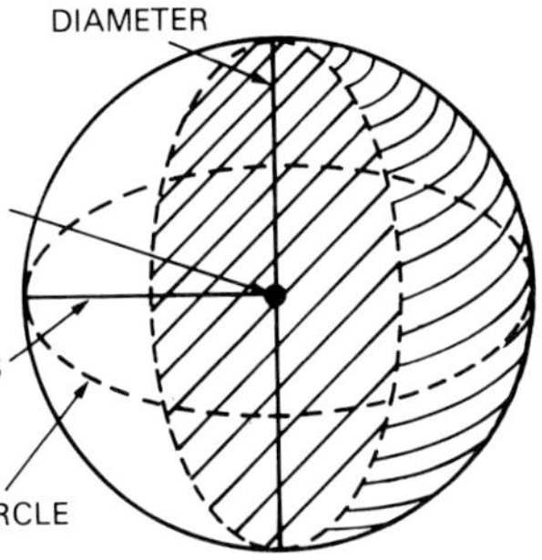

Fig. 6.43. Sphere

The **volume of a sphere** is equal to four-thirds of pi times the radius cubed:

$$V = \tfrac{4}{3}\pi r^3.$$

Example Problem: Find the volume of a sphere that has an 8-inch diameter.

Solution:

$V = \tfrac{4}{3}\pi r^3$, or $4 \times \pi \times r^3 \div 3$
$V = 4 \times 3.1416 \times 4 \times 4 \times 4 \div 3 = 268.0832 \text{ in.}^3$

A sphere is not a ruled surface; therefore, there is no way to "roll out" a sphere into a flat surface. Nevertheless, the area of a sphere can be determined by using the formula

$$S = \pi d^2$$

where

S = area of the sphere
d = diameter of the sphere.

Example Problem: Find the area of a sphere that has a 10-inch diameter.

Solution:

$S = \pi d^2$.
$S = 3.1416 \times 10 \times 10 = 314.16 \text{ in.}^2$

Table 6.4
Formulas for Finding Volumes

Figure	Formula	Meaning of Terms
Rectangular solid, cube	$V = lwh$	l = length w = width h = height
Cylinder	$V = Bh$, or $V = \pi r^2 h$	B = area of base h = height π = 3.1416 r = radius of base
Elliptical solid	$V = \pi abh$	a = one-half the major axis b = one-half the minor axis h = height π = 3.1416
Cone	$V = \frac{1}{3}(Bh)$, or $V = \frac{1}{3}(\pi r^2 h)$	B = area of base r = radius of base h = height, or altitude π = 3.1416
Pyramid	$V = \frac{1}{3}(Bh)$	B = area of base h = height, or altitude
Frustum of cone or pyramid	$V = \frac{1}{3}h(B + B' + \sqrt{BB'})$	h = height of frustum B = area of cone or pyramid base B' = area of section base
Sphere	$V = \frac{4}{3}\pi r^3$	r = radius of sphere

PRACTICE PROBLEMS

Refer to tables 6.3, 6.4, and 6.5 for help in solving these problems.

1. Cast iron weighs 450 pounds per cubic foot. What is the weight of the casting shown in the sketch below? ________________

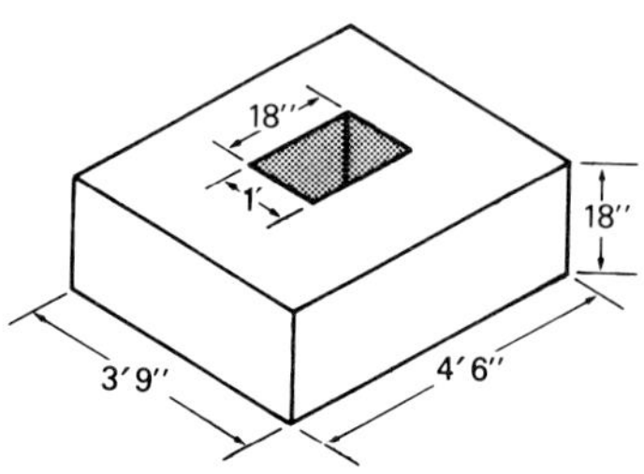

2. A ditch for a pipeline is to be 1,500 feet long, 12 inches wide, and 30 inches deep. How many cubic yards of dirt must be removed? ______

3. How many gallons of water will the rectangular trough shown below hold? ______

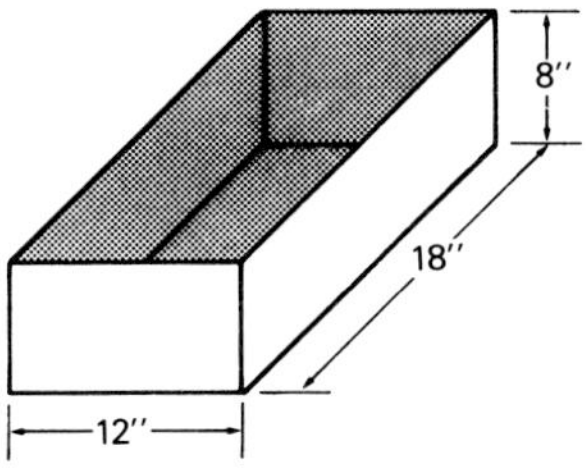

Table 6.5
Formulas for Finding Lateral/Total Surface Areas of Solid Figures

Figure	Formula	Meaning of Terms
Rectangular solid, cube	$T = 2A + 2A' + 2A''$	T = total area A = area of side 1 A' = area of side 2 A'' = area of side 3
Cylinder (right)	$L = 2\pi rh$ $L = hC$ $T = 2\pi r(r + h)$	L = lateral area r = radius of base h = altitude C = circumference of base π = 3.1416
Cone (right circular)	$L = ½sC$	s = slant height C = circumference
Pyramid (regular)	$L = ½sP$	s = slant height P = perimeter
Frustum of a cone (right circular)	$L = ½s(C + C')$ $T = L + B + B'$	s = slant height C = circumference of cone base C' = circumference of section base B = area of cone case B' = area of section base
Frustum of a pyramid (regular)	$L = ½s(P + P')$ $T = L + B + B'$	s = slant height P = perimeter of pyramid base P' = perimeter of section base B = area of pyramid base B' = area of section base

4. A cylindrical gasoline tank is 5 feet in diameter and 10 feet long. How many gallons of gasoline will it hold? ________________

5. A gauger gauges a cylindrical oil tank 15 feet in diameter and finds that the oil is 6½ feet deep. How many barrels of oil are in the tank? ________________

6. A single-cylinder pump has a stroke of 16 inches, and the cylinder is 12 inches in diameter. How many gallons of oil will the pump deliver at each stroke? ________________

7. A company makes metal signs out of ⅛-inch metal plate. The signs are cut in the form of an ellipse whose major axis is 3 feet and minor axis is 2 feet. If the plate weighs 489 pounds per cubic foot, what will be the weight of the sign? ________________

8. An engine foundation is to be poured 3 feet wide by 8 feet long by 5 inches deep. How many cubic feet of concrete will be needed? ________________

9. Calculate the number of cubic yards of concrete required to pour the footing represented in the sketch below. ________________

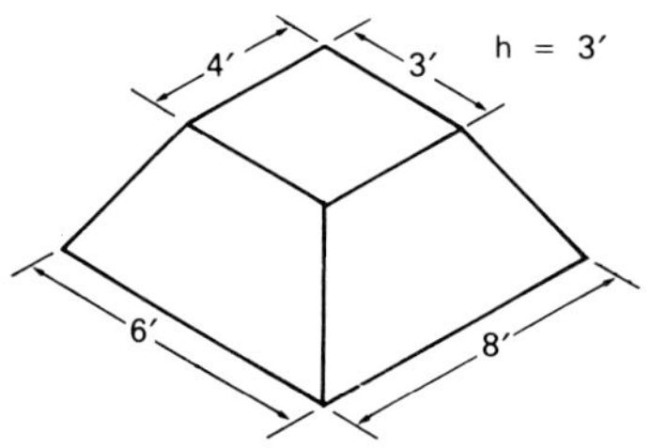

10. Find the volume of the oil heater shown in the sketch below. ________________

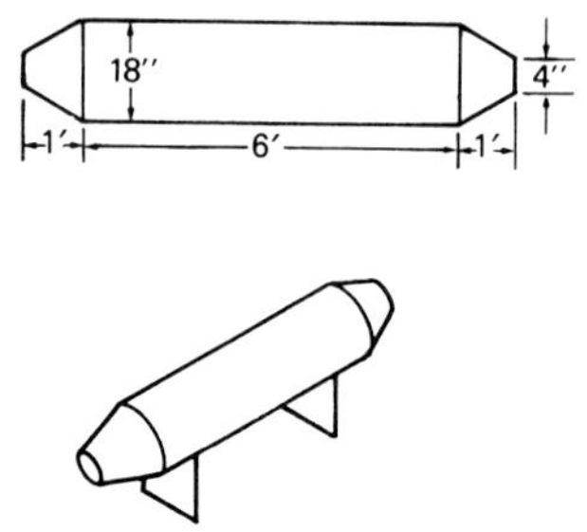

11. The casting shown in the sketch is made of brass that weighs 530 pounds per cubic foot. The casting has a hole that is 1½ inches in diameter through its center. Calculate its weight. ____________

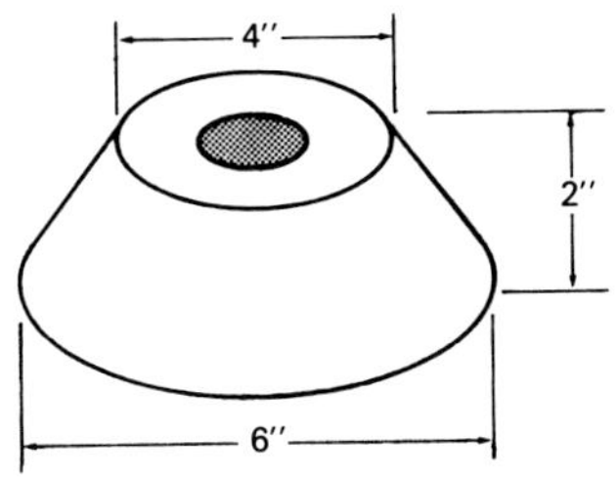

12. The exhaust hood shown below is connected to an exhaust fan that has a capacity of 500 cubic feet per minute.

a. What is the volume capacity of the hood? ____________

b. How many times per minute will the fan move that volume of air? ____________

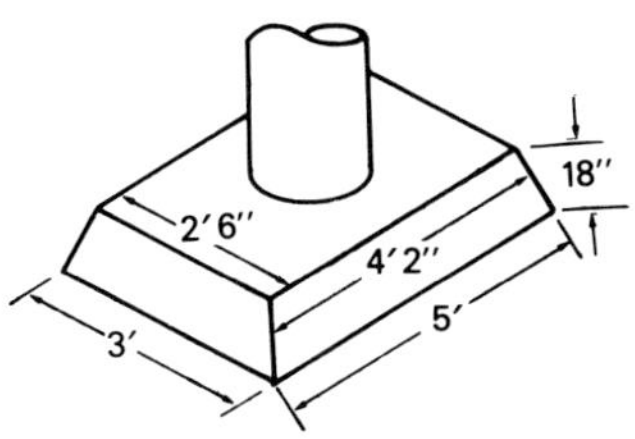

13. The compressor building whose floor is shown in the sketch below has a basement 8 feet deep. It is necessary to exhaust the air from the basement once each 3 minutes to prevent the accumulation of gases. How many fans, each with a capacity of 12,000 cubic feet per minute, are required? ____________

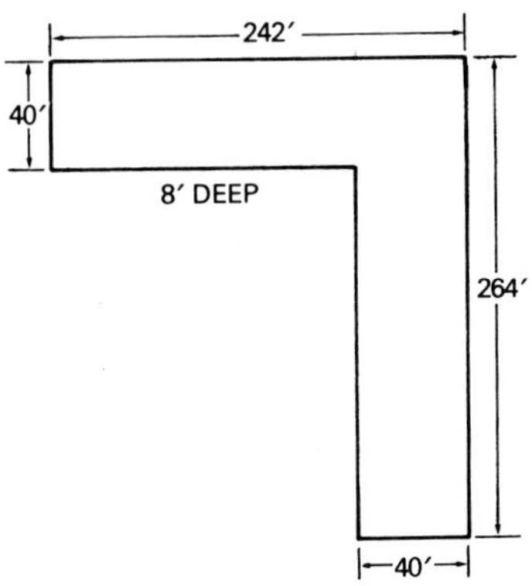

14. What will be the weight of the brass ball shown in the sketch if the brass of which it is cast weighs 534 pounds per cubic foot? ______________

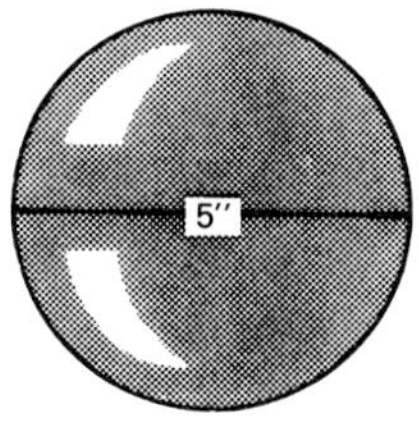

15. Calculate the lateral area of the doghouse sketched below. The door is a semicircle. *Note:* Divide into regular shapes, find the areas of each shape, and add together to determine the lateral area. The lateral area *does not* include the floor and roof. ______________

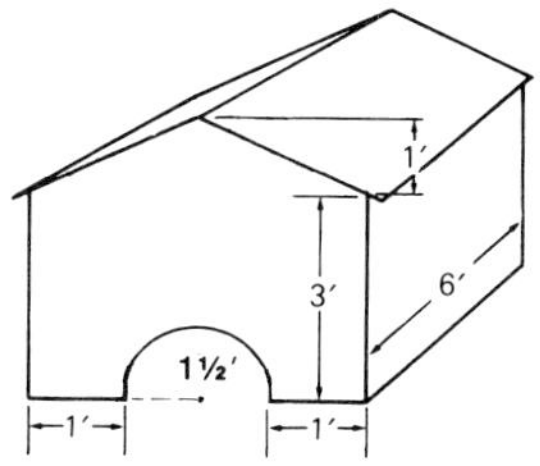

Geometric Constructions

Geometric construction involves laying out lines, angles, and curves in a regular pattern during design, so that the planned structures—such as buildings, pipes, and machinery—will fit together when built. The original layout is done on paper. Application on the job requires tape, rule, stakes, surveying instruments, or other tools and equipment. The same principles apply with chalk and stakes as with pencil and paper.

CONSTRUCTION TOOLS

Many mathematicians distinguish between *constructing* and *drawing* geometric figures. A compass and straightedge are the only tools used to *construct* figures, while many instruments are used in *drawing* them, the more common being a ruler, drawing board, T-square, and triangles. True constructions are accomplished without the benefit of a calibrated protractor for angles or a ruler for measuring lines.

A compass is a hinged instrument with a point on one arm and a drawing implement on the other. The angle between the two arms can be made wider or narrower as needed. The point of the compass is placed on a given point on the figure. The other end of the compass—the drawing end—is swung around without moving the point or the angle between the two arms of the compass. By swinging the drawing end of the compass, an arc is created. If the drawing end of the compass is swung all the way around, a circle is created.

The distance from the point end to the drawing end of the compass is called the *compass radius.* In figure 6.44, line *AB* is the compass radius. The curved line *CD* is the arc created by a compass with a radius of *AB.*

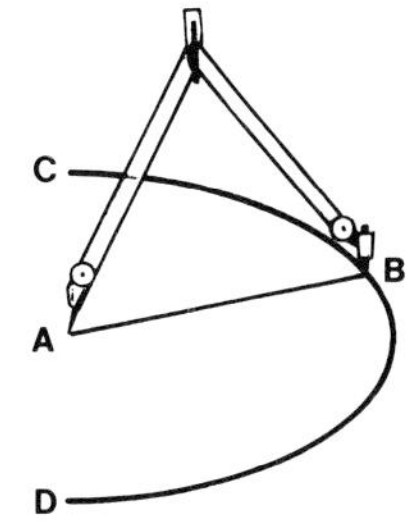

Fig. 6.44

BISECTING

Bisecting is dividing something in half with a line. A line, an arc, or an angle can be bisected by using only a compass and a straightedge.

Bisecting a line. Place the point of the compass at one end of the line (fig. 6.45), and swing the compass so that the drawing end lightly makes one arc above the line and another arc below the line. Then place the point of the compass at the other end of the line and repeat the previous procedure without changing the angle between the compass arms (fig. 6.46). The arcs above and below the line should cross at a single point. Use a compass radius that is greater than one-half the length of the line *AB* in order for the arcs to cross.

Using the straightedge, connect the points where the arcs cross with a line (fig. 6.47). Line *AB* now has been bisected by line *CD.* Line *CD* is called a *bisector* and is perpendicular to line *AB* (fig. 6.48).

Bisecting an arc. Use the same procedure as used for bisecting a line with points *A* and *B* occurring at the arc ends (fig. 6.49). The bisector is perpendicular to the chord joining the ends of the arc (chord *AB*).

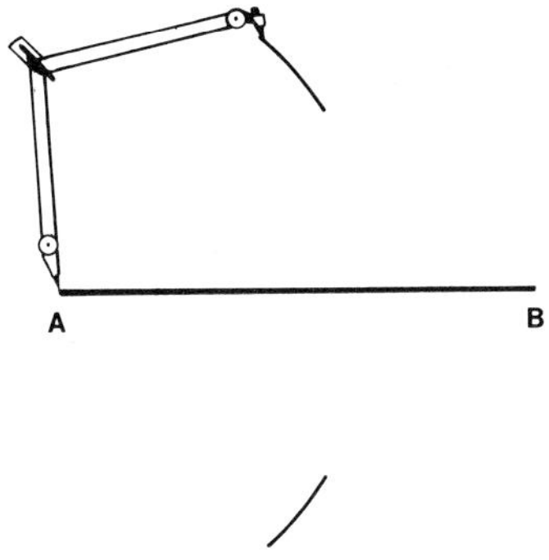

Fig. 6.45

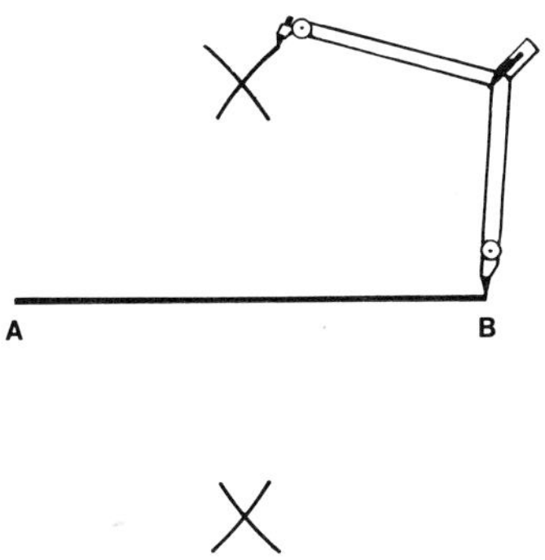

Fig. 6.46

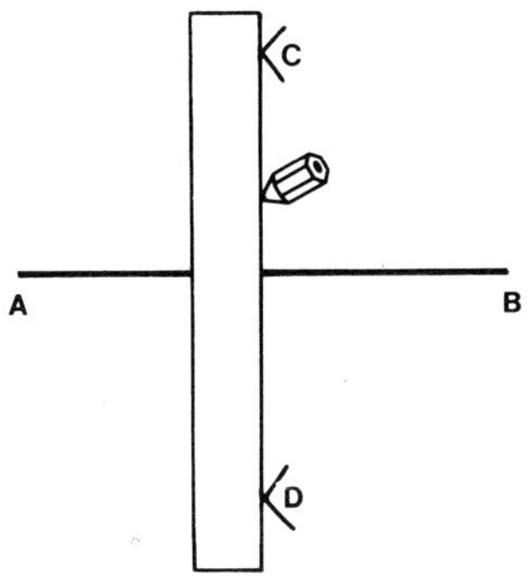

Fig. 6.47

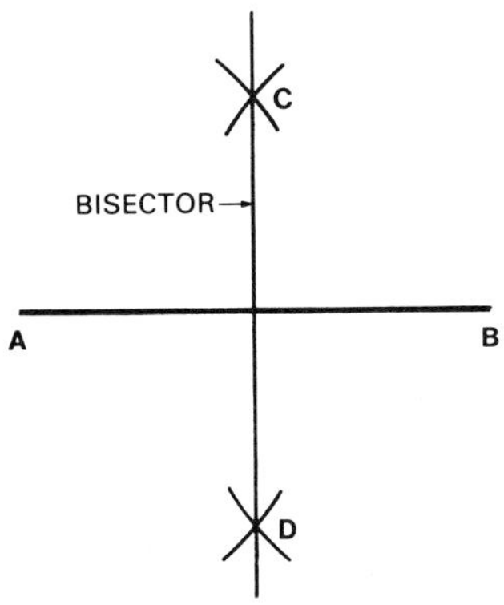

Fig. 6.48. Bisected line

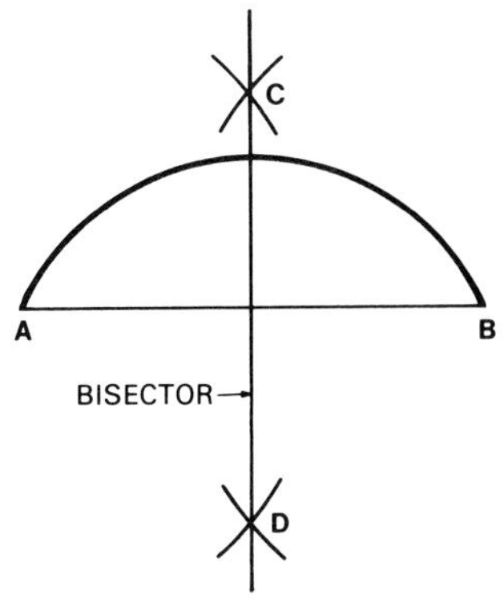

Fig. 6.49. Bisected arc

Bisecting an angle. Set the compass at a radius smaller than the angle sides, and then place the point of the compass at the angle vertex (*B* in figure 6.50). Swing the compass so that two arcs lightly cut across the sides of the angle (at *D* and *E*).

Now, without changing the compass radius, place the compass point at *E* and draw an arc through the middle of the angle. Move the compass point to *D* and draw an intersecting arc at *F* (fig. 6.51).

Using the straightedge, draw a line joining *B* and *F*. *Line BF* bisects angle *ABC* (fig. 6.52).

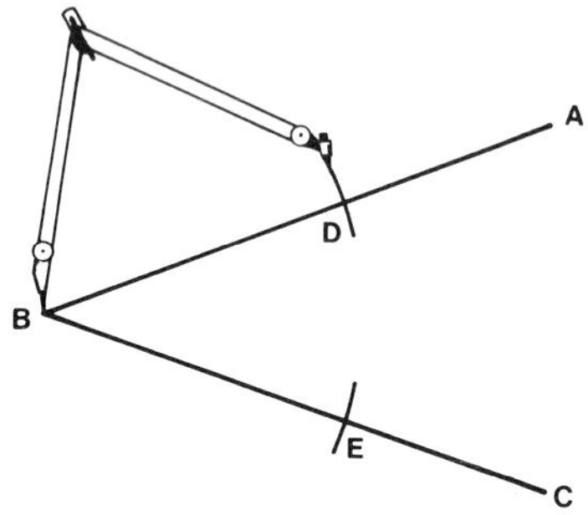

Fig. 6.50

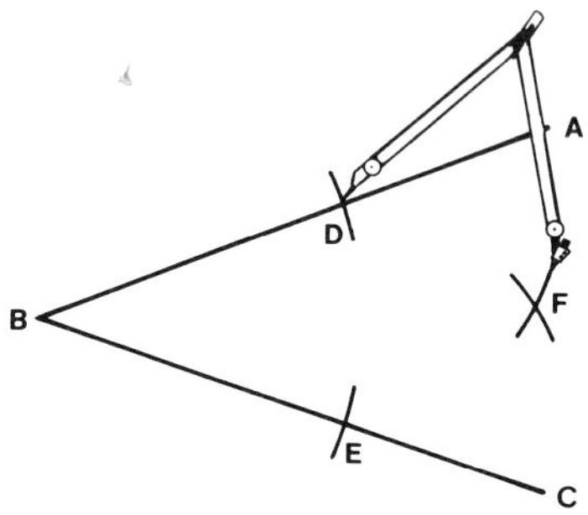

Fig. 6.51

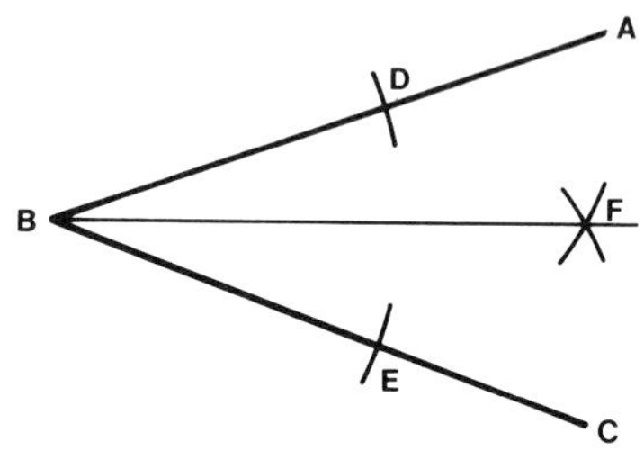

Fig. 6.52. Bisected angle

CONSTRUCTING FIGURES

The following figures can be constructed using only a compass and straightedge.

Constructing a circle passing through any three points not in a straight line. First, connect the three points (*A, B,* and *C* in figure 6.53) with straight lines, forming a triangle. Bisect any two sides of the triangle, such as *AB* and *BC.* The intersection of these two bisectors at point *D* will be the center of the circle. Place the compass point at *D,* adjust the compass radius to *DA, DB,* or *DC,* and draw the circle. It will pass through all three points (fig. 6.54).

Constructing an angle equal to a given angle. Draw a line *B′C′* to simulate line *BC* in angle *ABC* (fig. 6.55). Place the point of the compass at the vertex *B* of angle *ABC* and draw an arc that intersects sides *AB* at *D* and *BC* at *E* (fig. 6.56). Without changing the compass radius, place the compass point at *B′* and draw an arc that intersects *B′C′* at *E′*.

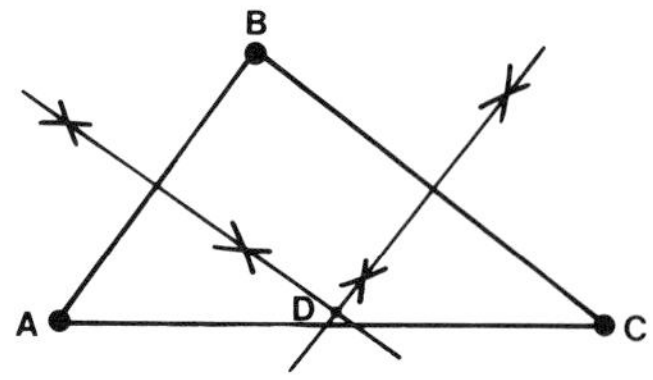

Fig. 6.53

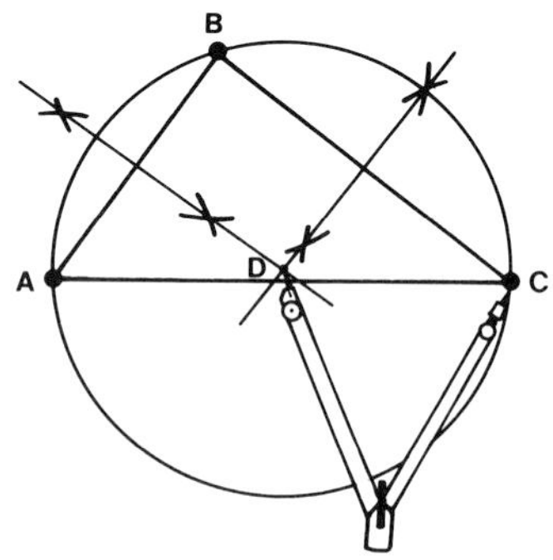

Fig. 6.54

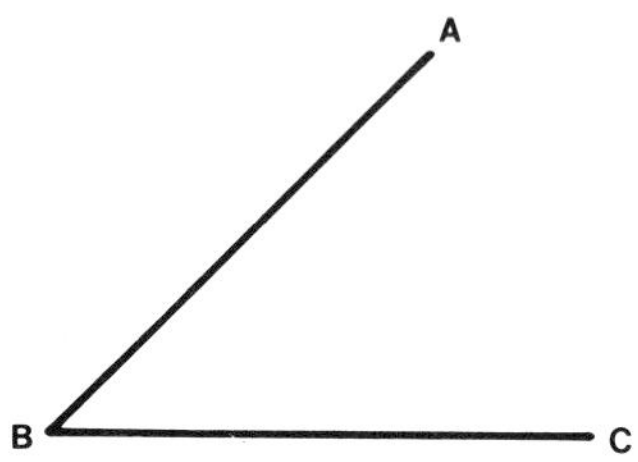

Fig. 6.55

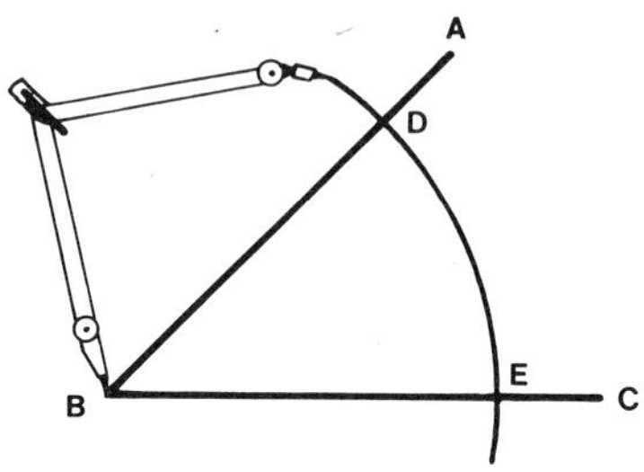

Fig. 6.56

Return to angle *ABC,* place the compass point at *E,* and adjust the compass angle so that the writing end will be on *D.* With this compass radius, and with *E′* as the center, make an arc that intersects the existing arc at *D′* (fig. 6.57). Using a straightedge, draw a line connecting points *B′* and *D′*. The resulting angle *A′B′C′* is equal to angle *ABC* (fig. 6.58).

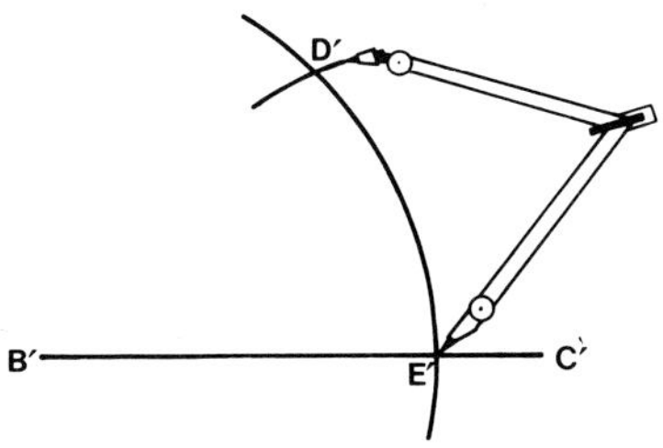

Fig. 6.57

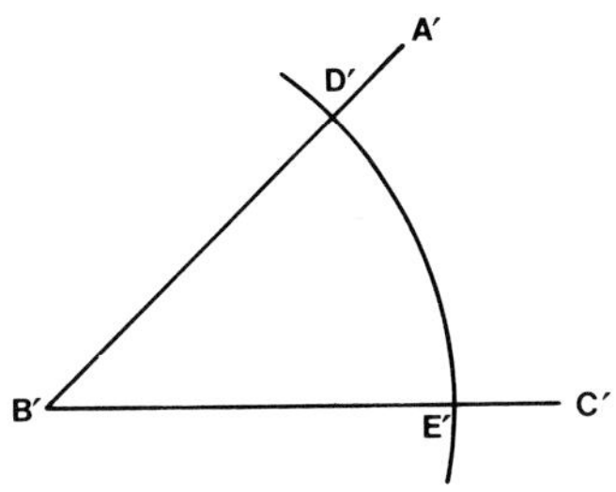

Fig. 6.58

Constructing a line through a given point that is parallel to a given line. To construct a line parallel to line *AB* that passes through point *P* in figure 6.59, draw line *DE* that connects point *P* with line *AB* at *C.* The next step is to draw another angle with the same measure as angle *ACD* located above point *P.* To do this, draw an arc (*FG*) through angle *ACD* and, without changing the compass radius, draw the same size arc using point *P* as the center (fig. 6.60).

With the compass, measure the distance between *F* and *G.* Using this compass radius and *H* as the center, draw a small arc that intersects the existing arc at point *I* (fig. 6.61). With the straightedge, draw line *JK* through points *I* and *P.* Line *JK* is parallel to line *AB* and passes through point *P* (fig. 6.62).

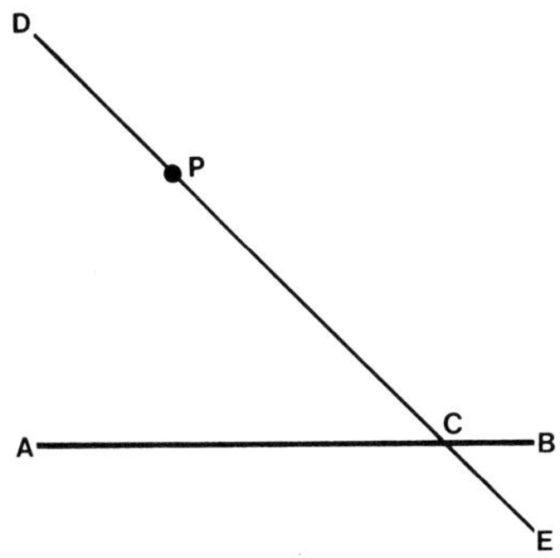

Fig. 6.59

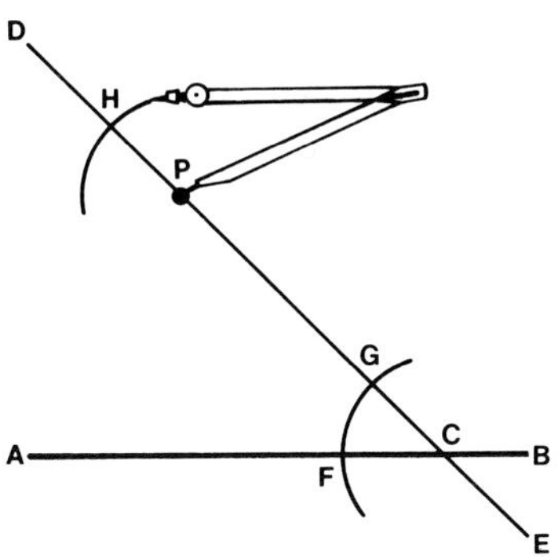

Fig. 6.60

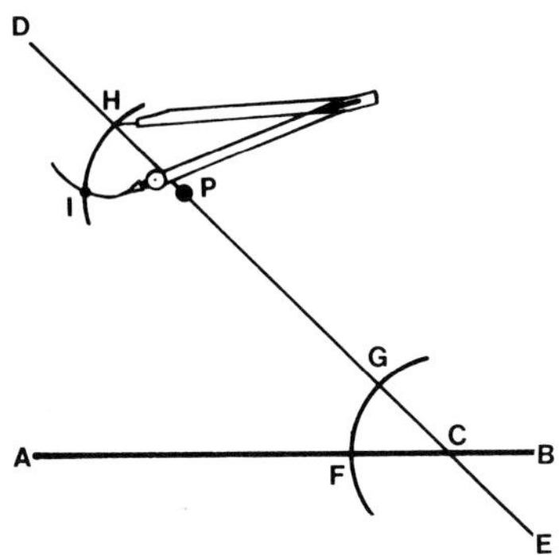

Fig. 6.61

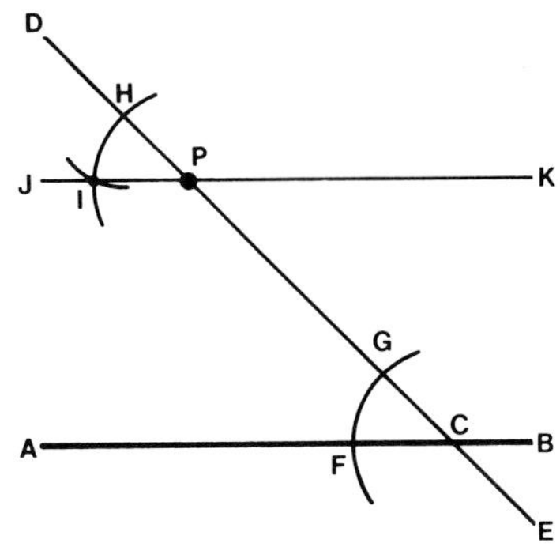

Fig. 6.62

Constructing a perpendicular to a line from a point on the line. To draw a line perpendicular to a given line *AB* at point *P*, place the compass point on *P*, and draw arcs intersecting the line at *C* and *D* on either side of *P* (fig. 6.63). Now enlarge the compass radius slightly, and with the compass point on *C*, draw a large arc above the line. Without changing the compass radius, move the compass point to *D* and draw another arc that intersects the first arc at *E* (fig. 6.64). Using a straightedge, draw a line connecting *E* and *P*. Line *EP* is perpendicular to line *AB* (fig. 6.65).

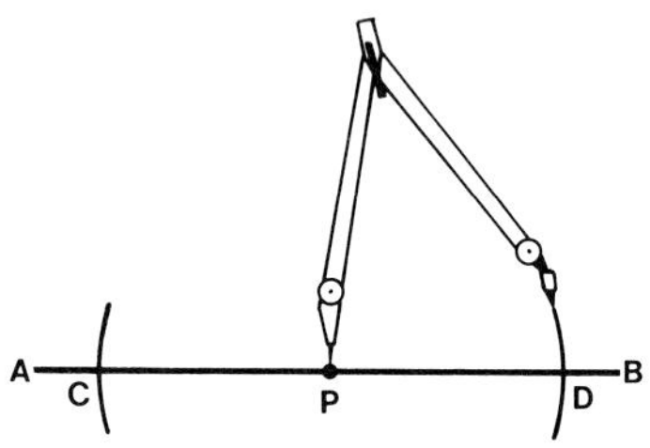

Fig. 6.63

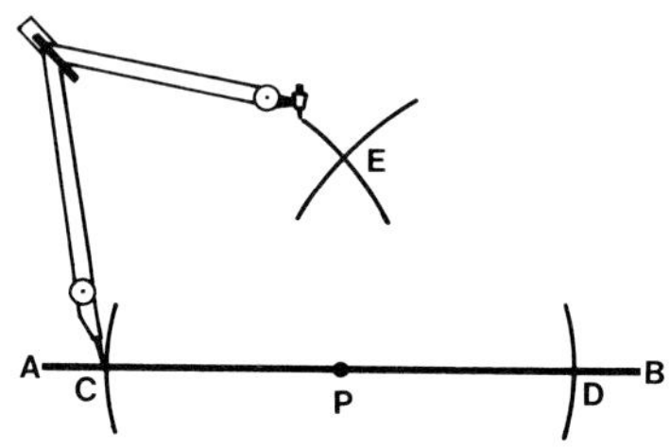

Fig. 6.64

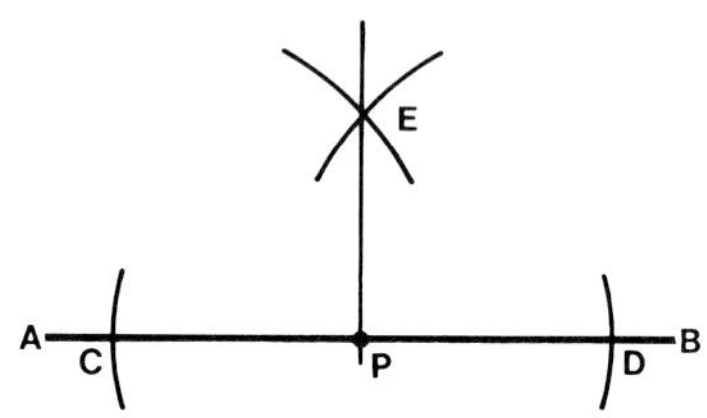

Fig. 6.65

Constructing a perpendicular to a line from a point near the end of the line. Another method can be used to draw a perpendicular line when the desired point is very close to the end of the line. To draw a line perpendicular to line *AB* at point *P* (fig. 6.66), select a second point *C* somewhere above line *AB*. Place the compass point at *C* and the compass drawing end at *P*. With the compass radius at this setting, draw a circle.

This circle intersects line *AB* at points *P* and *D*. With a straightedge, draw a line connecting *C* at the circle's center with the intersection of circle and line at *D*. Extend this line all the way to the other side of the circle through *C* to intersect the circle at *E* (fig. 6.67).

The new line *DCE* is the circle's diameter. Point *E* is directly above the original point *P* on line *AB*. Using a straightedge, draw a line connecting points *E* and *P*. Line *EP* is perpendicular to the given line *AB* (fig. 6.68).

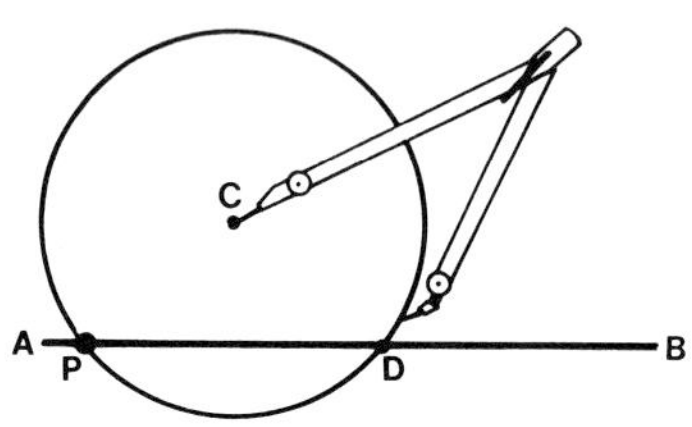

Fig. 6.66

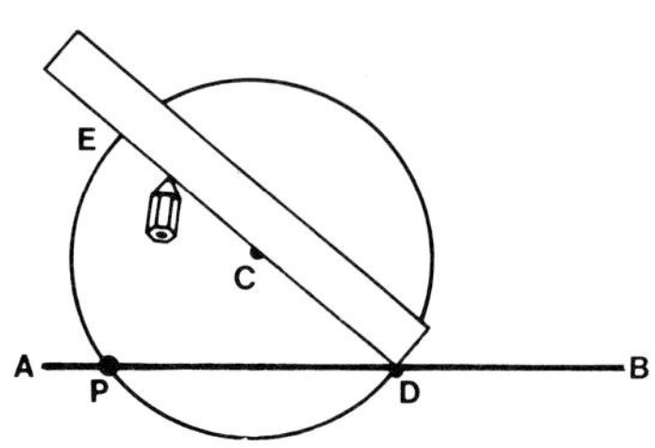

Fig. 6.67

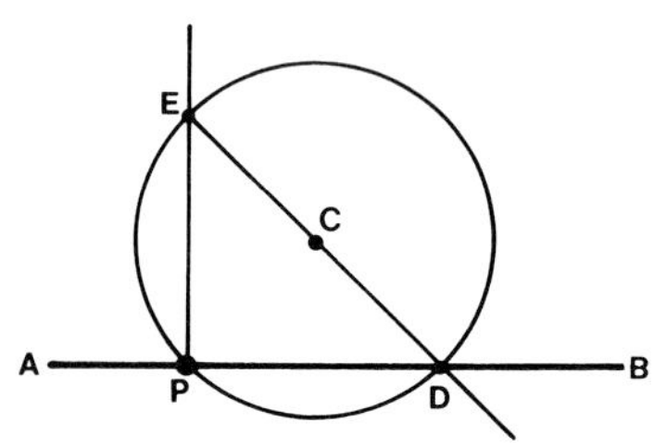

Fig. 6.68

Constructing a perpendicular to a line from a point not on the line. To draw a perpendicular to a line *AB* from a point *P* that is not on the line, put the compass point on *P* and set the compass at an angle wide enough to draw an arc that intersects line *AB* in two places (points *C* and *D* in fig. 6.69). Then set the compass radius so that it is larger than half the distance between points *C* and *D* (fig. 6.70). Placing the compass point on *C,* draw an arc below the line. Without changing the compass radius, move the compass point to *D* and draw another arc below the line. The arcs below the line should intersect at *E* (fig. 6.71). With the straightedge, draw a line connecting point *P* above the line with point *E* below the line. The new line *PE* is perpendicular to line *AB* (fig. 6.72).

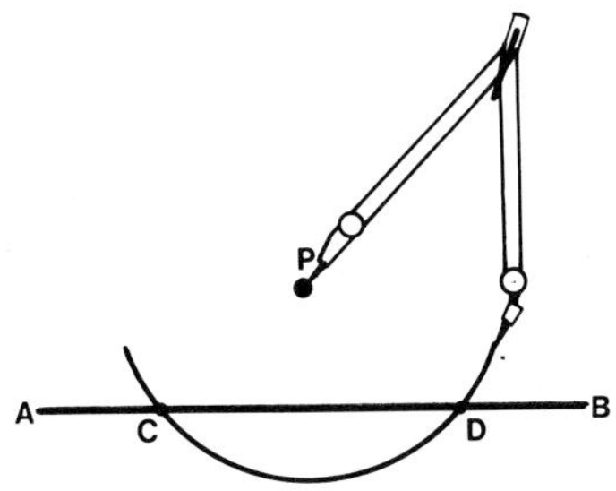

Fig. 6.69

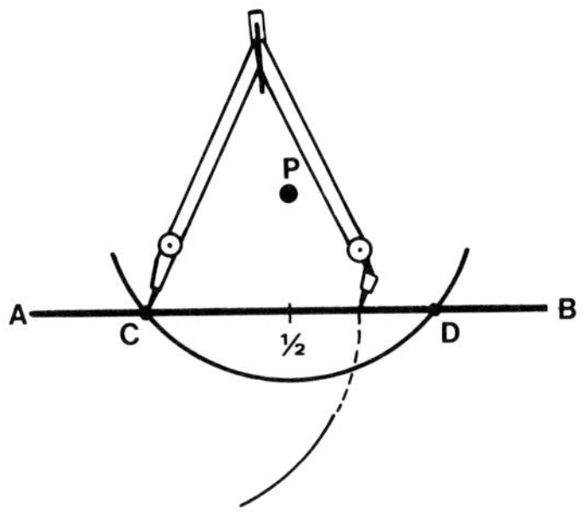

Fig. 6.70

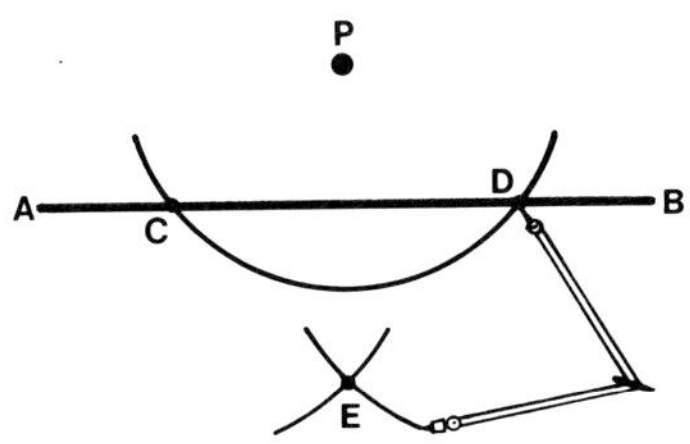

Fig. 6.71

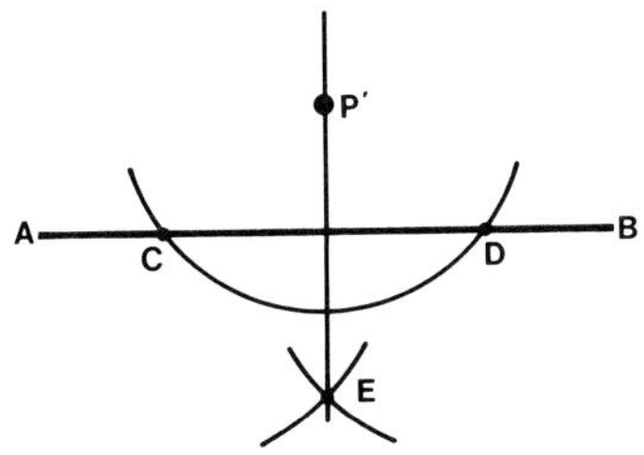

Fig. 6.72

Dividing a given line into any number of equal parts. To divide line *AB* into five equal parts, draw line *AC* at any convenient angle, starting at point *A* (fig. 6.73). From *A,* mark off on line *AC* five equal divisions of any length using dividers or a compass. From point 5 on *AC,* draw a line to *B* (fig. 6.74). Then construct lines from 1, 2, 3, and 4 that are parallel to line 5*B.* (Use the procedure for drawing parallel lines given earlier.) These lines divide line *AB* into five equal parts.

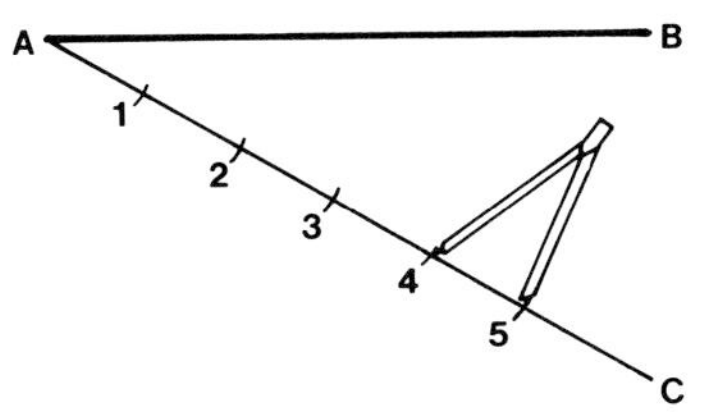

Fig. 6.73

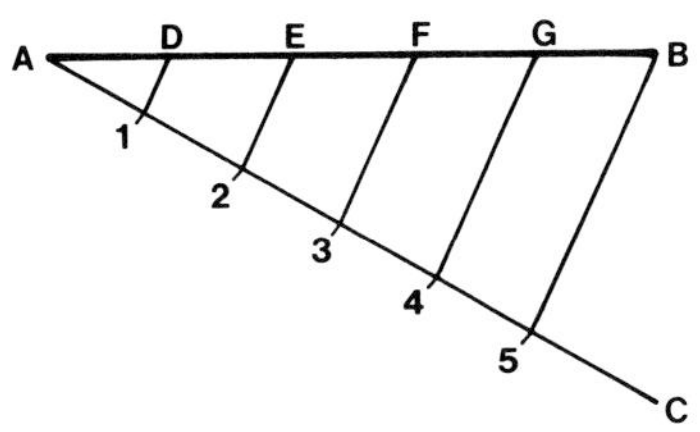

Fig. 6.74

Laying out a half circle into six equal parts. Draw line *CD* perpendicular to the semicircle diameter *AB* that intersects at *D* (fig. 6.75). (Use the method of constructing perpendiculars described earlier.) Using the distance between *C* and *D* as a compass radius, construct arcs intersecting the semicircle to the left and right of the perpendicular at points *E* and *F.* Without changing the compass radius, move the compass point to *A* and construct an arc intersecting the semicircle at point *G* (fig. 6.76). Move the compass point to *B* and draw another arc intersecting the semicircle at point *H.* The circumference is now divided into six equal parts at points *E, G, C, H,* and *F.* Connecting these points to point *D* with a straightedge divides the semicircle into six equal parts.

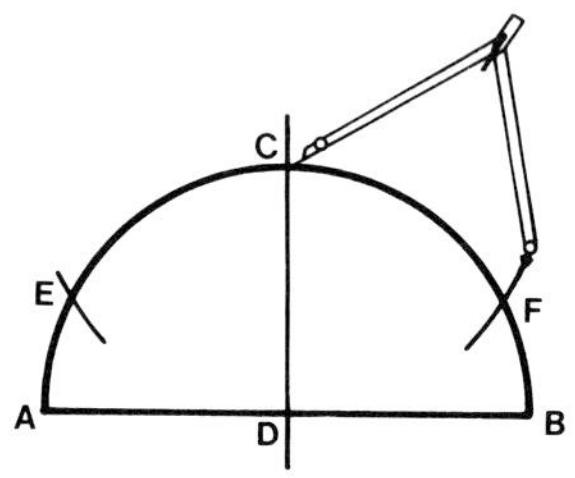

Fig. 6.75

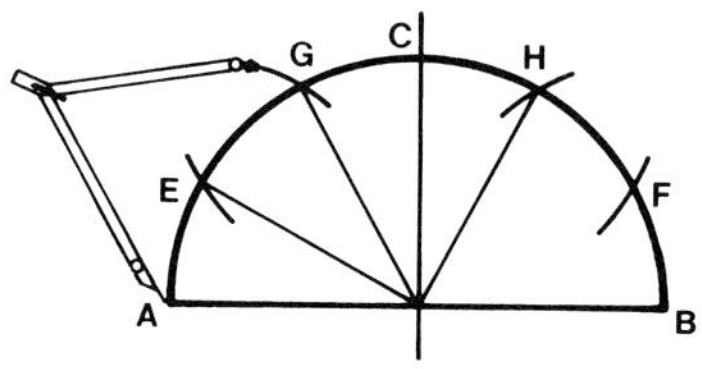

Fig. 6.76

CONSTRUCTING TANGENTS

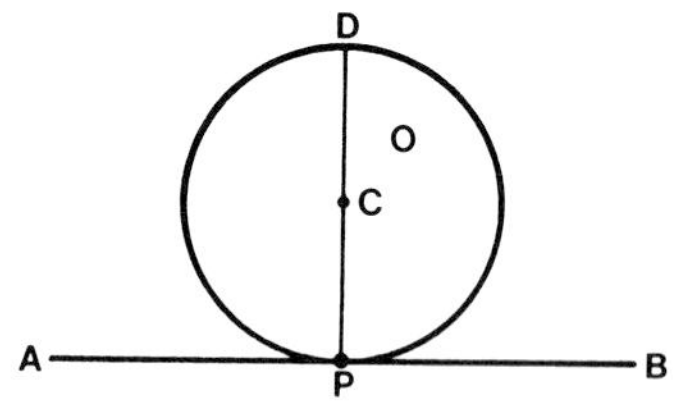

Fig. 6.77. Circle tangent to line

The word *tangent* means touching, and a line that touches a circle at only one point is said to be tangent to the circle. Likewise, the circle is said to be tangent to the line. The point where the two touch is known as the *point of contact,* or *point of tangency.* Several theorems apply to lines and circles that are tangent to each other (fig. 6.77):

1. A straight line perpendicular to a radius at its outer extremity is tangent to the circle. (*AB* is perpendicular to *CP* and therefore is tangent to circle *O*.)
2. A line perpendicular to a tangent at its point of contact passes through the center of the circle. (Line *PD* is perpendicular to *AB* and therefore passes through center *C* of circle *O*.)
3. A line from the center of a circle and perpendicular to a tangent passes through the point of contact. (Line *CP* is perpendicular to *AB* and therefore passes through point *P*.)

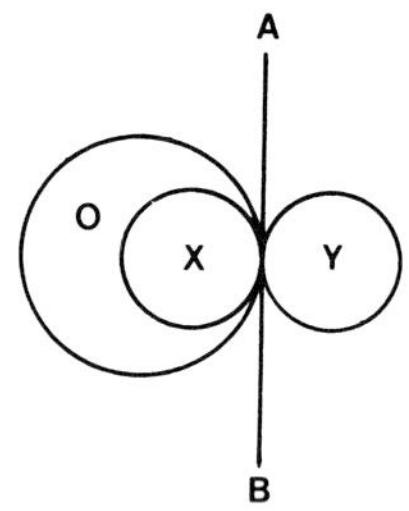

Fig. 6.78. Tangent circles

Two circles are said to be tangent to each other if they are both tangent to the same line at the same point, either internally or externally. In figure 6.78, circles *O* and *X* are tangent internally, and circles *O* and *Y* are tangent externally. Both are tangent to line *AB*. If two circles are tangent, a straight line drawn through their centers (known as the *line of centers*) passes through their point of contact. In other words, line *DE* drawn through *C* and *C′* passes through *P* (fig. 6.79). Also, line *DE* is perpendicular to line *AB*.

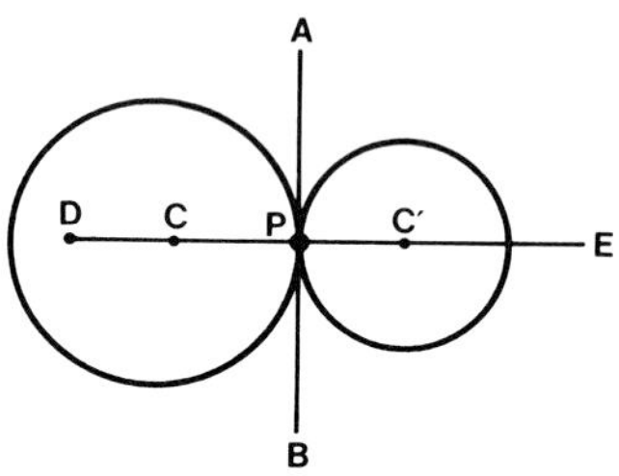

Fig. 6.79. Line of centers

Solid figures may also be tangent to each other. A line can be tangent to a sphere if it touches the sphere at only one point. Like circles and lines, spheres and planes have similar theorems (fig. 6.80):

1. A plane perpendicular to a diameter of a sphere at one end is tangent to the sphere. (Plane *X* is perpendicular to diameter *DP* and therefore is tangent to sphere *O*.)
2. If a plane is tangent to a sphere at the end of a diameter, it is perpendicular to the diameter. (If plane *X* is tangent to sphere *O* at *P*, then it is perpendicular to diameter *DP*.)
3. If a sphere is tangent to a plane, it is tangent to any line in the plane passing through the point of contact. (Sphere *O* is tangent to plane *X* and therefore tangent to line *AB*, line *YZ*, and all lines passing through point *P*.)

Like tangent circles, tangent spheres have their line of centers pass through the point of contact. These facts should help in constructing the following tangents.

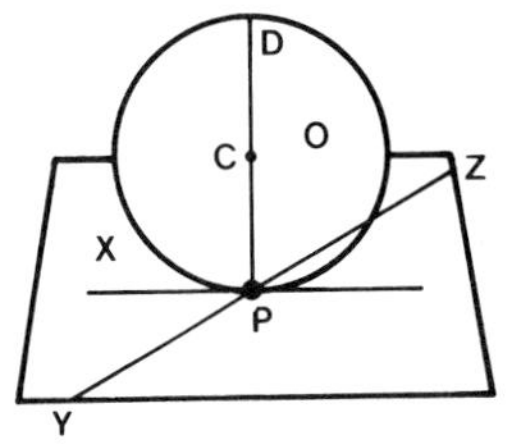

Fig. 6.80. Sphere tangent to plane

Constructing a line tangent to a circle from a point on the circle. Using a straightedge, draw a line connecting the circle's center *C* with the given point *P* on the circle (fig. 6.81). Then construct a line perpendicular to line *CP* at *P*, using the procedure for constructing perpendiculars given earlier. Line *AB* is tangent to circle *O*.

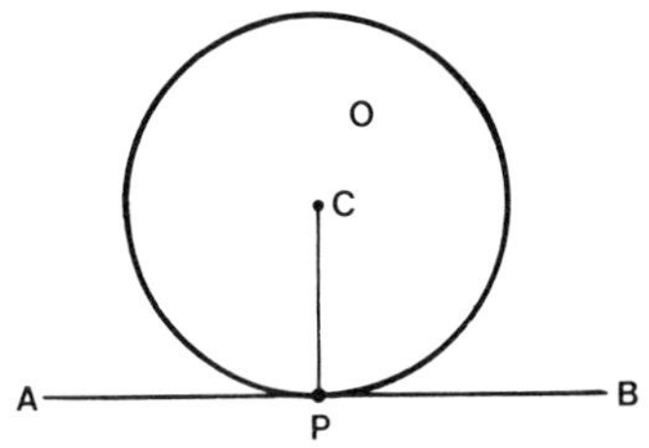

Fig. 6.81

Constructing lines tangent to a circle from a point outside the circle. With the straightedge, connect the circle's center *C* with the given point *P* outside the circle (fig. 6.82). Using the bisecting method described, bisect line *CP*, producing point *A* that lies halfway between the end points of line *CP*. Measure line *CA* with the compass, and with this compass radius and the compass point at *A*, draw a second circle, which should go through the center *C* of circle *O* and point *P* outside the circle. The new circle intersects circle *O* at *B* and *D* (fig. 6.83). Now, with a straightedge, draw lines *BP* and *DP*, which are tangent to circle *O* (fig. 6.84).

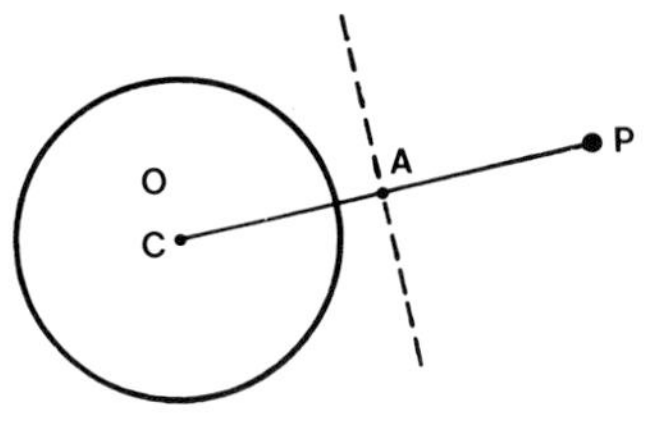

Fig. 6.82

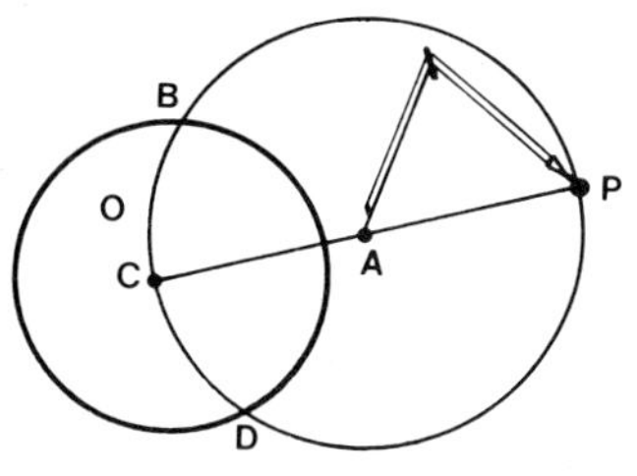

Fig. 6.83

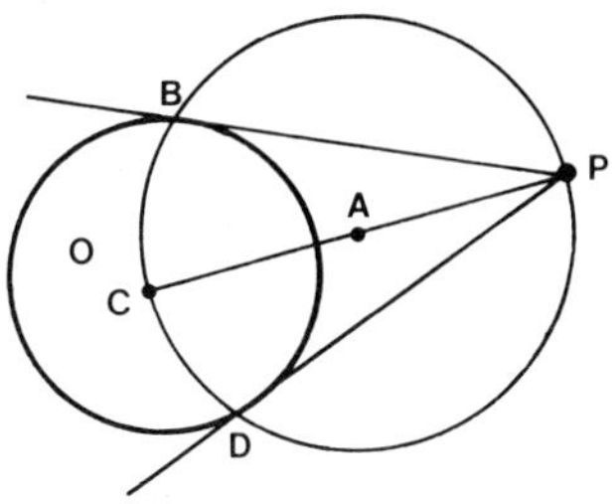

Fig. 6.84

Constructing a circular arc of given radius, tangent to two non-parallel lines. Draw perpendicular lines *EF* and *EG* to the two given lines *AB* and *CD* using the method described earlier (fig. 6.85). Set the compass radius equal to *XY*. With the compass point at the intersection of lines *AB* and *FE*, draw an arc that intersects the perpendicular line *FE* at point *K*. With the same compass radius, repeat the procedure using lines *CD* and *EG*, creating an arc at point *L*.

Now using the method described for constructing parallel lines, draw a line *IH* parallel to *AB* at point *K* and line *IJ* parallel to *DC* at point *L*. The new lines intersect at point *I*. With the compass radius set the same and with point *I* as the center, construct an arc that touches the original lines *AB* and *CD* at *M* and *N* (fig. 6.86). This arc is tangent to the non-parallel lines *AB* and *CD* at points *M* and *N*.

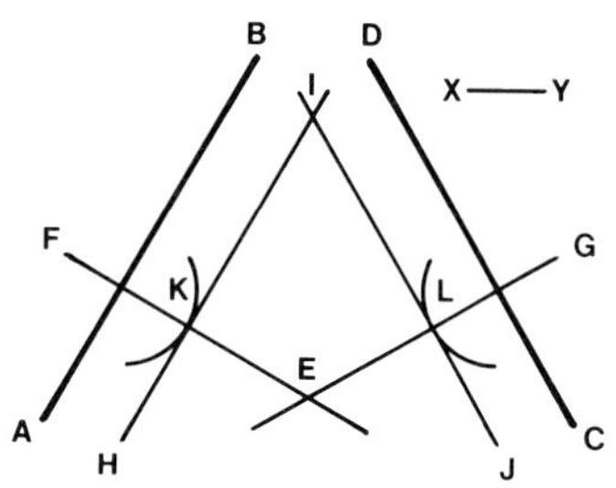

Fig. 6.85

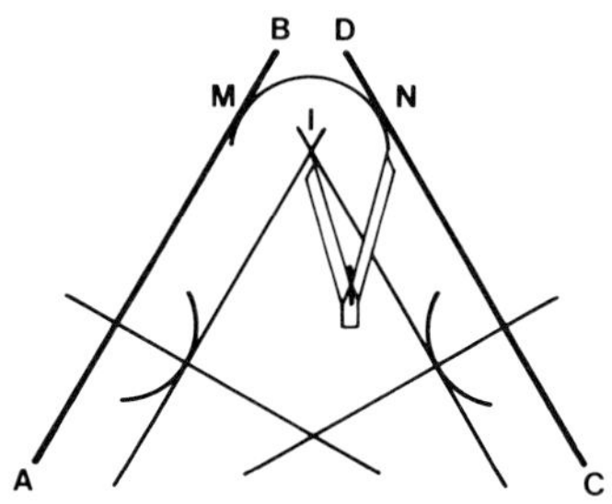

Fig. 6.86

DRAWING PROCEDURES

Geometric figures can be *drawn* instead of *constructed* by using a drawing board, a T-square, and various-sized triangles. The more commonly used triangles are the 45°-45° and the 30°-60°, meaning that the triangles have 45° and 45° angles or 30° and 60° angles opposite the right angle (fig. 6.87).

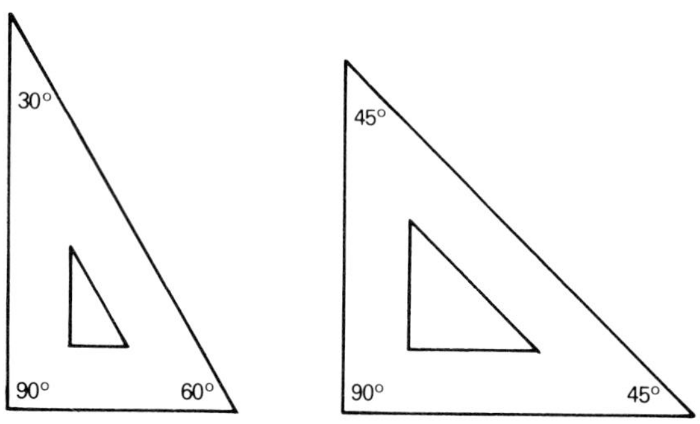

Fig. 6.87. Drawing triangles

The proper use of triangles and T-square for drawing lines is illustrated in figure 6.88. Arrows indicate the direction in which to draw the lines.

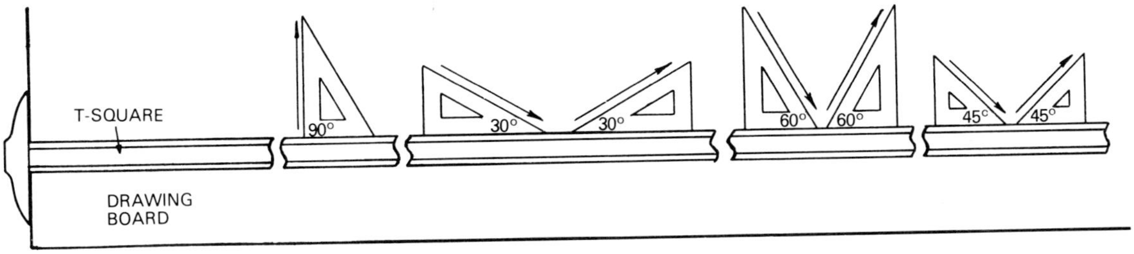

Fig. 6.88. Proper drawing strokes

Drawing a line parallel to a given line. To draw a line parallel to the given line *AB* (fig. 6.89), place a triangle in position so that one edge coincides with line *AB.* Place the T-square blade against the other edge of the triangle. Hold the T-square firmly in place, slide triangle to desired new position, and draw parallel line *CD* (fig. 6.90).

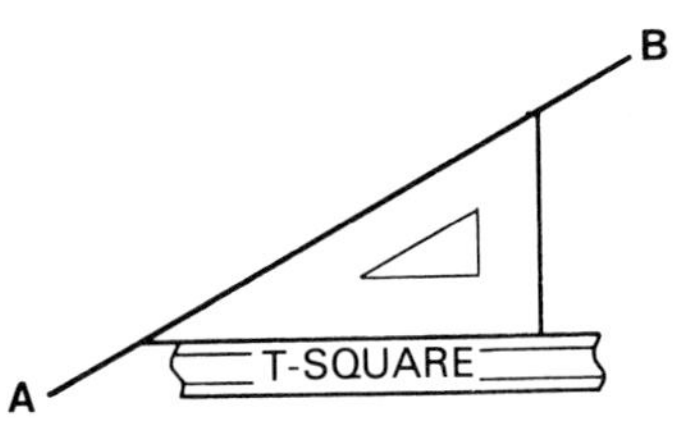

Fig. 6.89

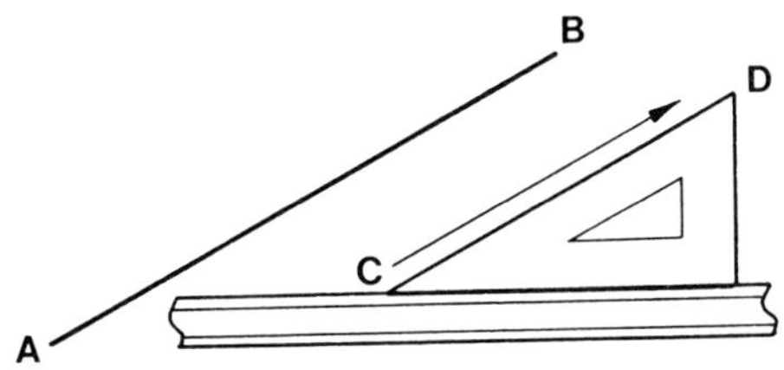

Fig. 6.90

Drawing a line perpendicular to a given line. To draw a line perpendicular to the given line *AB,* place a 30°-60° triangle so that its hypotenuse coincides with the line (fig. 6.91). Hold the T-square firmly in place and turn the triangle about its 90° angle to the position shown in figure 6.92, then draw line *CD.* Line *CD* is perpendicular to line *AB.*

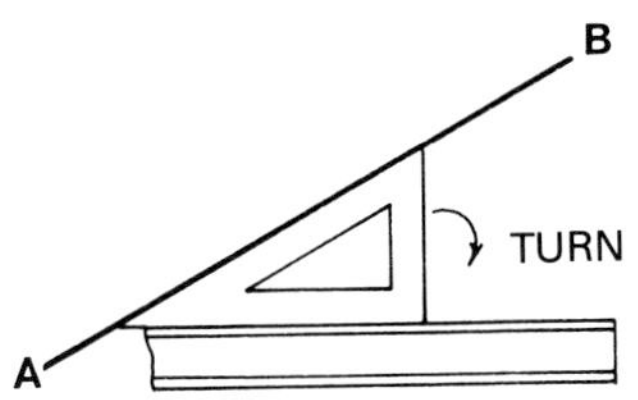

Fig. 6.91

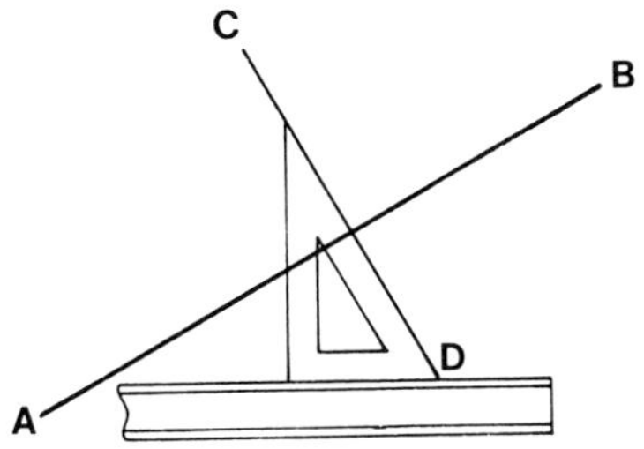

Fig. 6.92

Dividing a circle into equal parts. By using both triangles, a circle can be divided into as many as twenty-four equal parts, each part having a 15° angle. Figure 6.93 shows the various steps involved in achieving the final "sunburst" pattern. Arrows indicate the direction in which to draw the lines.

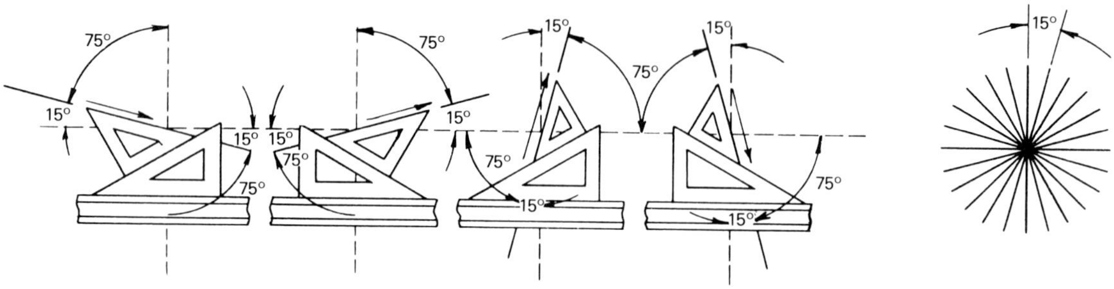

Fig. 6.93

PRACTICE PROBLEMS

(*Note: Since these are construction exercises, no answers are given.*)

1. Erect a perpendicular to a line that is 2¾ inches long at a point that is 1⅛ inches from the end of the line.

2. Erect a perpendicular to a line 3⅜ inches long at one end of the line.

3. Draw a line across the boiler plate in the sketch from point B and perpendicular to the line AB. Erect perpendiculars to line BC at each end of the line.

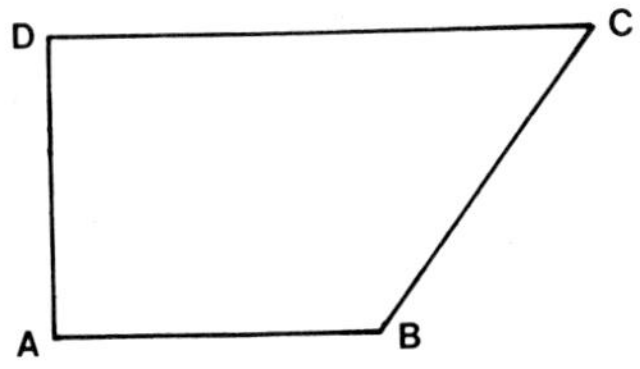

4. Bisect a line that is 3½ inches long.

5. Lay out an angle of 75° with a protractor. Then bisect this angle using only a compass and a straightedge. Show all construction lines.

6. Divide a line 2⅝ inches long into seven equal parts.

7. Draw a line AB. Draw a perpendicular line CD at point D on line AB. Angle CDB is a 90° angle. Construct an angle of 45° by bisecting angle CDB.

8. The drawing below is of a pipe that is to be installed in the corner of a building. With a compass and straightedge, determine graphically the length of the pipe joining the two 45° ells.

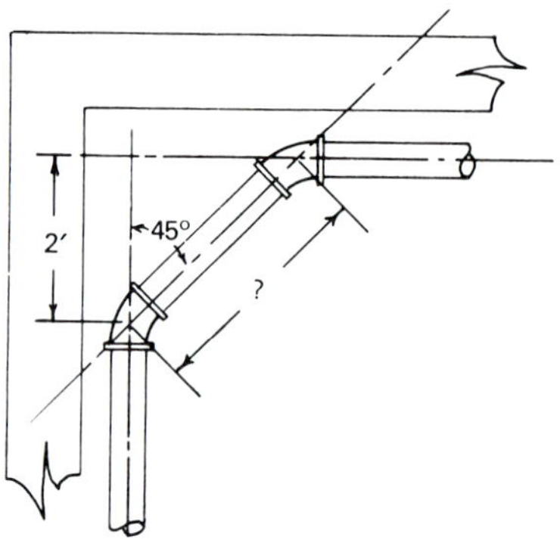

9. Draw a triangle. Then construct another triangle so that two angles and one side will be equal respectively to two angles and the side of the first.

10. Draw a circle. Construct a tangent to the circle at a point on the circle.

SELF-TEST
6. Practical Geometry

Multiply each question/problem answered correctly by five to arrive at your percentage of competency.

Use tables 6.1–6.5 as a reference for solving these problems. Using conversion factors given in Appendix C-1 will alter answers somewhat.

1. Find the area of a circle with a radius of 2 millimetres.

   ____

2. Three holes have been drilled into a small metal plate. The diameter of each hole is ⅜ of an inch. What is the total area of the holes?

   ____

3. How many gallons will the tank shown below hold?

   ____

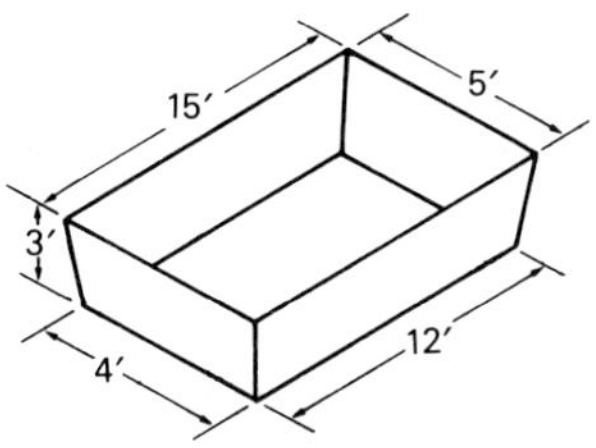

4. How many square inches of braking surface are there on the clutch disk shown in the sketch below?

   ____

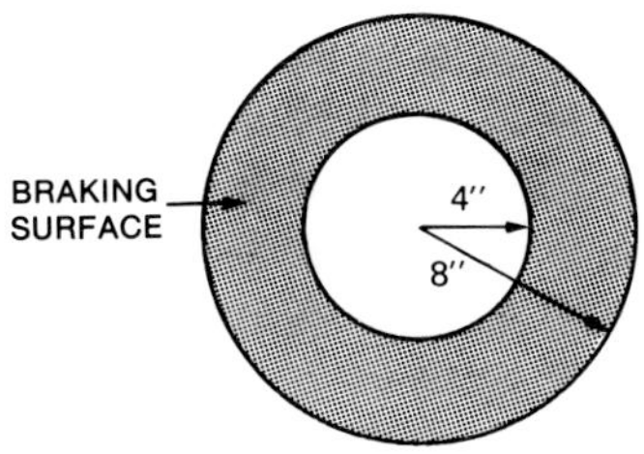

5. Find the volume of a sphere 14 inches in diameter.

   ____

6. How many barrels will the cylindrical tank in the sketch below hold?

____

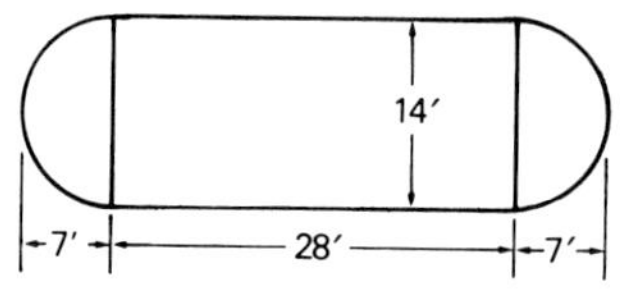

7. How many square inches are contained in a triangular gusset plate if its base is 8 inches and its height is 4½ inches?

____

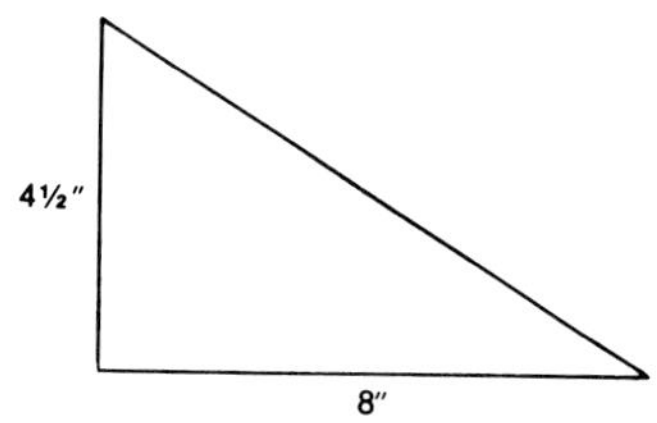

8. A farm shaped in the form of a parallelogram is cut in two by a highway as shown below. How many acres are there in the triangle formed at the right of the road?

____

How many acres are there on the left of the road?

____

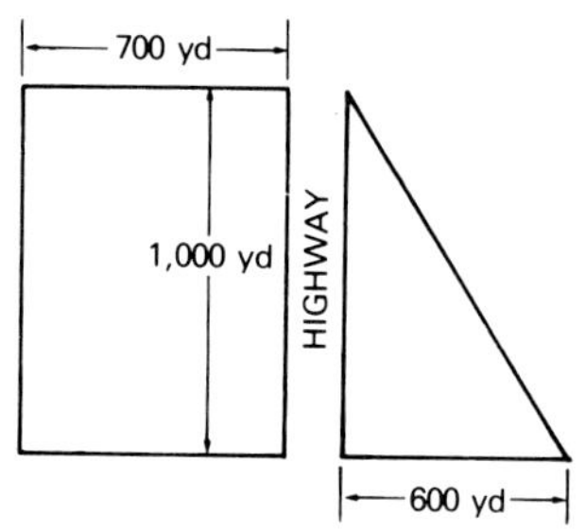

9. How much paint will it take for the two gable ends of the roof shown in the sketch if 1 gallon will cover 350 square feet?

____

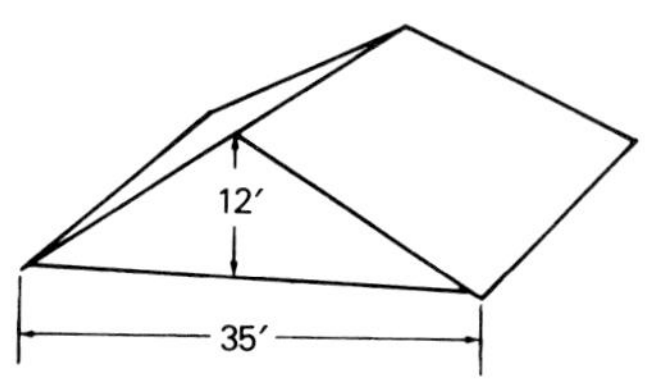

10. A water tower is constructed as shown in the sketch. How many gallons of water are needed to fill the tank to the top of the conical section? ________________

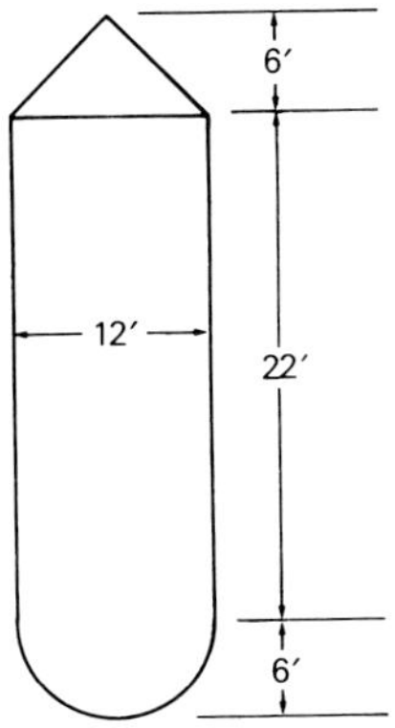

11. The tank in the sketch below is in the shape of a half ellipse. How many gallons will this tank hold? ________________

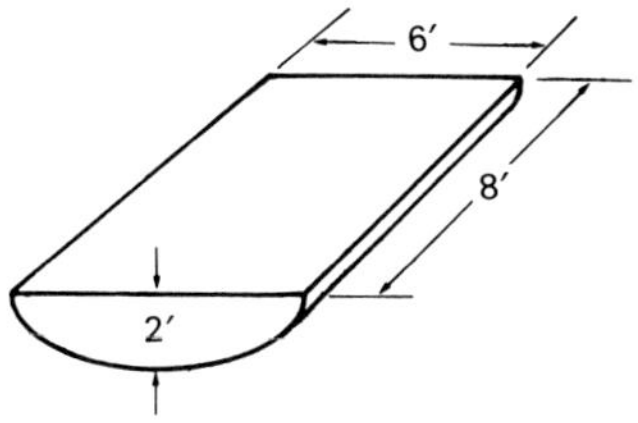

12. In the five right triangles shown below, solve for x.

a. 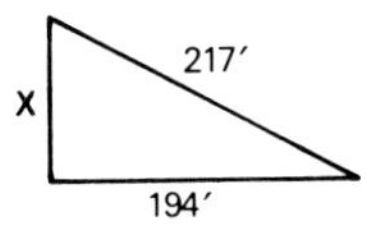 ________________

b. 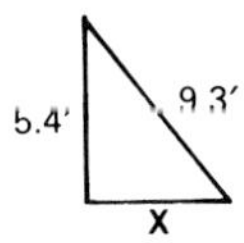 ________________

c. 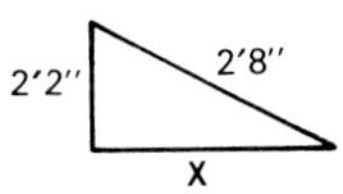 ________________

d. 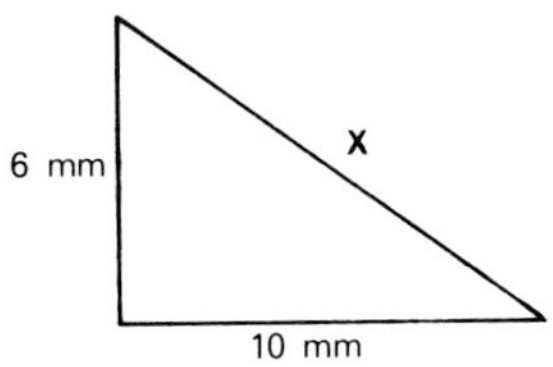 ________________

e. 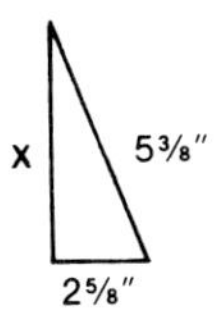 ________________

13. How many bolts will be required for the joint shown on the sketch of the tank if the bolt holes are spaced 3″ apart? The outside diameter of the tank is 10′6″.

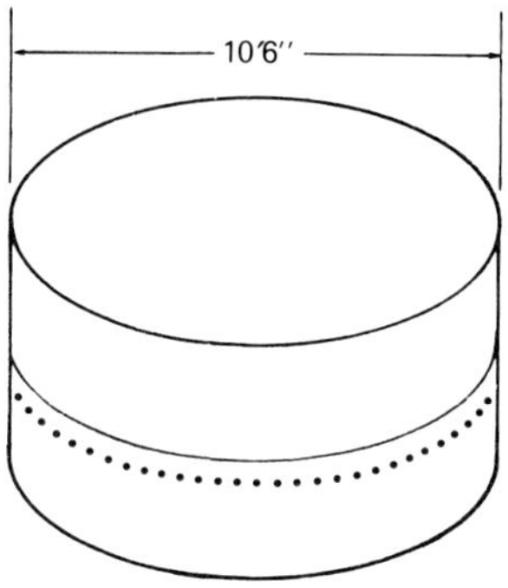

14. Find the number of square feet in a floor that is 28′6″ by 18′9″.

15. If 1 gallon of paint will cover 400 square feet, how many gallons will be needed to paint the walls of a room that is 18 feet long, 12 feet wide, and 8 feet high?

16. Find the area of a trapezoid whose parallel sides are 10 inches and 20 inches and whose height is 7 inches.

17. Find the number of cubic feet in nine pieces of plastic that are 2 inches thick, 6 inches wide, and 3 feet long. Round off to two decimal places.

18. The diameter of a truck tire fully inflated is 30 inches. How many revolutions per minute will the wheel make if the car travels at the rate of 40 miles per hour?

19. Find the lateral area of a cone whose slant height is 6 inches and whose base diameter is 4 inches.

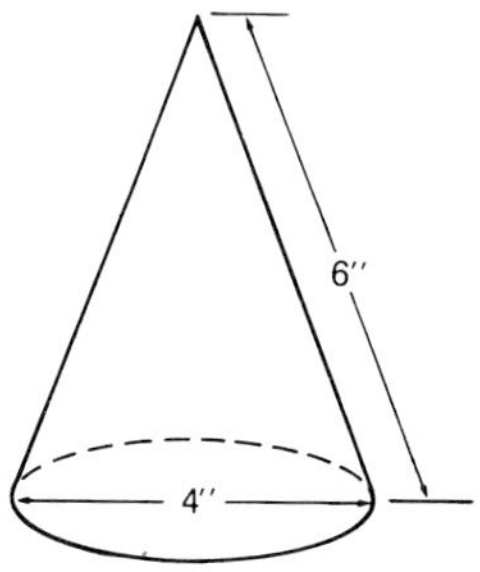

20. Find the volume of the cone frustum shown below.

____

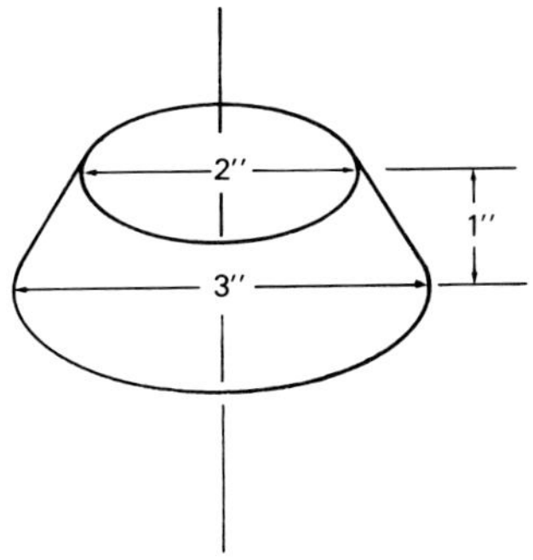

7
TRIGONOMETRY

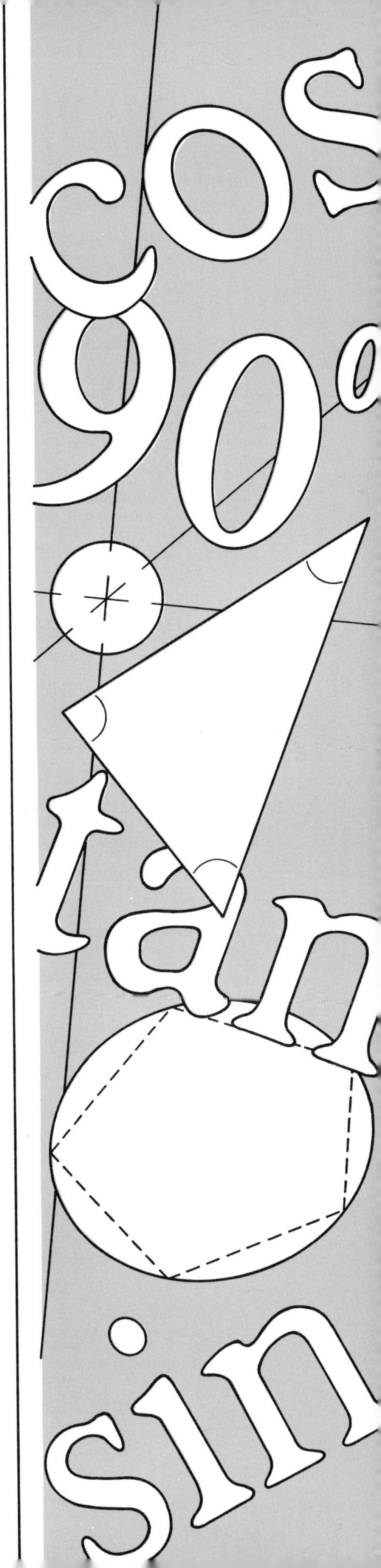

Objectives

Upon completion of chapter seven, the student will be able to—

1. Explain the principles of right triangle trigonometry.
2. Define the three most common trigonometric functions, or ratios: sine, cosine, and tangent.
3. Use trigonometric functions to find the unknown measurement of one side of a right triangle when the other two sides or one acute angle and one side are known.
4. Use a table of trigonometric functions or a scientific calculator to quickly solve for missing values in triangles.
5. Name the reciprocals of the sine, cosine, and tangent.
6. Determine which trigonometric function to use when solving problems involving triangles.
7. Solve for unknown measurements of oblique triangles by using trigonometric formulas.
8. Find areas of triangles by using trigonometric functions.
9. Apply trigonometric formulas to everyday situations by constructing triangles to represent the situations.
10. Solve problems requiring the use of inverse functions.

Trigonometry is the study of the properties of triangles and their use in solving problems. It is that branch of mathematics that deals with the relations of the sides and angles of triangles and with methods of computing unknown data from data given or known. For example, a surveyor can compute a distance that he cannot physically measure by determining angles and lengths with his surveying instrument. Then, using these known values and trigonometric ratios, he can find the unknown measurement.

To understand the basis for trigonometry, consider the two triangles in figure 7.1. Even though triangle *A'B'C'* is much larger than triangle *ABC,* the ratios of the sides and angles are the same. From these ratios come the principles of trigonometry.

The right angle triangle is the basis for all trigonometry calculations because the relations between its angles and sides are the simplest and best known. Oblique triangles—those having no right angles—can also be dealt with by using trigonometric formulas.

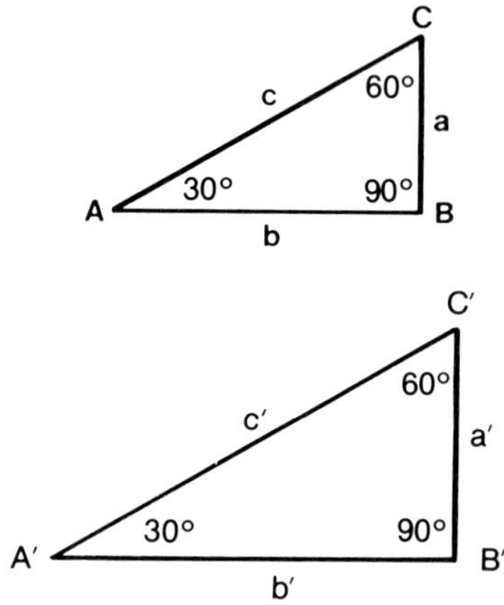

Fig. 7.1. Regardless of triangle size, ratios of sides and angles remain the same.

Right Triangle Trigonometry

Right triangle trigonometry, as the name implies, deals with the solution of problems involving triangles in which one angle is a right angle. When values of two of the sides or one angle and one side of a right triangle are known, the remaining angles and sides may be computed indirectly.

Angles are measured from an arbitrary radial line of a circle. The units of measure for angles are degrees, minutes, and seconds. There are 360 degrees (360°) in a circle, 60 minutes (60′) in 1 degree, and 60 seconds (60″) in 1 minute. A quadrant is one-fourth of a circle, or 90 degrees, and a 90-degree angle is known as a right angle. A right triangle has one right angle (90°) and two acute angles, whose sum is 90 degrees, making a total of 180 degrees in a triangle.

UNDERSTANDING TRIGONOMETRIC FUNCTIONS

Trigonometric functions, or ratios, involve two sides and an acute angle of a right triangle. Sides of a triangle are generally referred to as the *opposite,* the *adjacent,* and the *hypotenuse.* Looking at figure 7.2, side c is the hypotenuse and is opposite the right angle *C.* Side *a* is opposite angle *A,* and side *b* is opposite angle *B.* Side *a* is adjacent to angle *B,* and side *b* is adjacent to angle *A.* Of course, side c is also adjacent to angles *A* and *B,* but *the side opposite the right angle is always called the hypotenuse.* Understanding this terminology is important because the functions are defined in terms of these sides. The three most commonly used trigonometric functions are the sine (sin), cosine (cos), and tangent (tan).

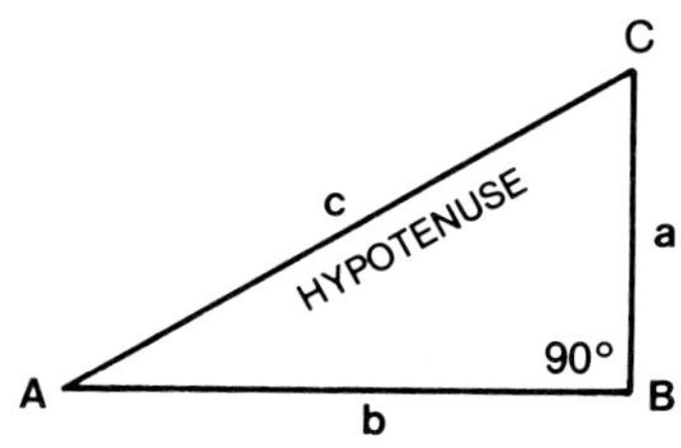

Fig. 7.2. Sides of a triangle

Sine. The sine of an acute angle in a right triangle is found by dividing the side opposite the angle by the hypotenuse:

$$\text{sine} = \frac{\text{opposite side}}{\text{hypotenuse}} .$$

In figure 7.2, the sine of angle A (referred to as sin A) would be

$$\sin A = a/c,$$

and the sine of angle B would be

$$\sin B = b/c.$$

It should be noted that in any right triangle, the ratio is constant and changes only when the acute angle involved changes. As the acute angle increases, the value of the sine increases. The sine of a small angle is small, and the sine of a large angle is large. As an angle approaches 90°, its sine becomes larger and is equal to 1 at 90° (as shown in table 7.1). The values of the trigonometric ratios are based on acute angles, those lying between 0° and 90°. However, the ratios exist for angles greater than 90°. For angles between 90° and 180° (obtuse angles), the sine values repeat, being 1 at 90° and decreasing to 0 at 180°. For angles between 180° and 360°, the values of the sines will be negative; for example, sin 270° = −1.00.

The table of natural trigonometric functions (table 7.1) gives the sine and other functions for any angle from 0° to 90° and is useful in solving problems. Scientific calculators that have the trigonometric functions can also be used to find the ratios of an angle. To find the sine or other function of an angle with a calculator, the manufacturer's instructions should be followed; however, most calculators require entering the angle, then pressing the function key.

Example Problem: Find the numerical value of sin 26°.

Solution: Using table 7.1, locate 26° in the left-hand column. Then read the answer found under the "Sine" column: 0.4384. Or using a calculator, press "26 SIN" and get 0.4383711.

Example Problem: Find the sine of angle B in the right triangle shown in figure 7.3.

Solution: Since the sum of angles A and B equals 90°, to find angle B, subtract:

$$90° - 39° = 51°.$$

Then, from table 7.1 or using the calculator, sin 51° = 0.7771.

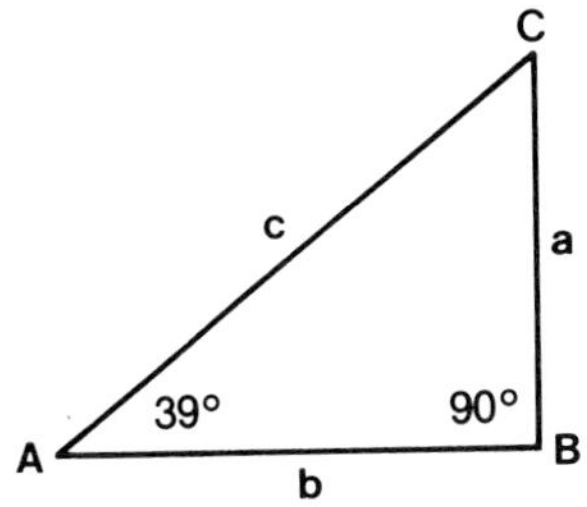

Fig. 7.3

Table 7.1
Natural Trigonometric Functions

Angle	Sine	Cosine	Tangent	Angle	Sine	Cosine	Tangent
1°	0.0175	0.9998	0.0175	46°	0.7193	0.6947	1.0355
2°	0.0349	0.9994	0.0349	47°	0.7314	0.6820	1.0724
3°	0.0523	0.9986	0.0524	48°	0.7431	0.6691	1.1106
4°	0.0698	0.9976	0.0699	49°	0.7547	0.6561	1.1504
5°	0.0872	0.9962	0.0875	50°	0.7660	0.6428	1.1918
6°	0.1045	0.9945	0.1051	51°	0.7771	0.6293	1.2349
7°	0.1219	0.9925	0.1228	52°	0.7880	0.6157	1.2799
8°	0.1392	0.9903	0.1405	53°	0.7986	0.6018	1.3270
9°	0.1564	0.9877	0.1584	54°	0.8090	0.5878	1.3764
10°	0.1736	0.9848	0.1763	55°	0.8192	0.5736	1.4281
11°	0.1908	0.9816	0.1944	56°	0.8290	0.5592	1.4826
12°	0.2079	0.9781	0.2126	57°	0.8387	0.5446	1.5399
13°	0.2250	0.9744	0.2309	58°	0.8480	0.5299	1.6003
14°	0.2419	0.9703	0.2493	59°	0.8572	0.5150	1.6643
15°	0.2588	0.9659	0.2479	60°	0.8660	0.5000	1.7321
16°	0.2756	0.9613	0.2867	61°	0.8746	0.4848	1.8040
17°	0.2924	0.9563	0.3057	62°	0.8829	0.4695	1.8807
18°	0.3090	0.9511	0.3249	63°	0.8910	0.4540	1.9626
19°	0.3256	0.9455	0.3443	64°	0.8988	0.4384	2.0503
20°	0.3420	0.9397	0.3640	65°	0.9063	0.4226	2.1445
21°	0.3584	0.9336	0.3839	66°	0.9135	0.4067	2.2460
22°	0.3746	0.9272	0.4040	67°	0.9205	0.3907	2.3559
23°	0.3907	0.9205	0.4245	68°	0.9272	0.3746	2.4751
24°	0.4067	0.9135	0.4452	69°	0.9336	0.3584	2.6051
25°	0.4226	0.9063	0.4663	70°	0.9397	0.3420	2.7475
26°	0.4384	0.8988	0.4877	71°	0.9455	0.3256	2.9042
27°	0.4540	0.8910	0.5095	72°	0.9511	0.3090	3.0777
28°	0.4695	0.8829	0.5317	73°	0.9563	0.2924	3.2709
29°	0.4848	0.8746	0.5543	74°	0.9613	0.2756	3.4874
30°	0.5000	0.8660	0.5774	75°	0.9659	0.2588	3.7321
31°	0.5150	0.8572	0.6009	76°	0.9703	0.2419	4.0108
32°	0.5299	0.8480	0.6249	77°	0.9744	0.2250	4.3315
33°	0.5446	0.8387	0.6494	78°	0.9781	0.2079	4.7046
34°	0.5592	0.8290	0.6745	79°	0.9816	0.1908	5.1446
35°	0.5736	0.8192	0.7002	80°	0.9848	0.1736	5.6713
36°	0.5878	0.8090	0.7265	81°	0.9877	0.1564	6.3138
37°	0.6018	0.7986	0.7536	82°	0.9903	0.1392	7.1154
38°	0.6157	0.7880	0.7813	83°	0.9925	0.1219	8.1443
39°	0.6293	0.7771	0.8098	84°	0.9945	0.1045	9.5144
40°	0.6428	0.7660	0.8391	85°	0.9962	0.0872	11.4301
41°	0.6561	0.7547	0.8693	86°	0.9976	0.0698	14.3007
42°	0.6691	0.7431	0.9004	87°	0.9986	0.0523	19.0811
43°	0.6820	0.7314	0.9325	88°	0.9994	0.0349	28.6363
44°	0.6947	0.7193	0.9657	89°	0.9998	0.0175	57.2900
45°	0.7071	0.7071	1.0000	90°	1.0000	0.0000	∞

Cosine. The cosine of an acute angle in a right triangle is obtained by dividing the side adjacent to the angle by the hypotenuse:

$$\text{cosine} = \frac{\text{adjacent side}}{\text{hypotenuse}} .$$

Again, referring to figure 7.2, the cosines of angles A and B would be

$$\cos A = b/c$$
$$\cos B = a/c.$$

Note in table 7.1 that as the angle increases, the cosine decreases. The cosine of an acute angle is never equal to or greater than 1 because 1 would be 0°. The cosine of a small angle is large.

Practice Problem: Find the numerical value of cos 16°.

Solution: Table 7.1 shows that the cosine of 16° is 0.9613. Entering "16 COS" on the calculator gives 0.9612617.

Practice Problem: In the right triangle in figure 7.4, find b if $c = 213$ feet and angle $A = 41°$.

Solution: Using the cosine function, transpose and substitute known values.

$$\cos A = b/c$$
$$b = c \times \cos A$$
$$b = 213 \text{ ft} \times \cos 41° = 213 \times 0.6561$$
$$b = 139.75 \text{ ft.}$$

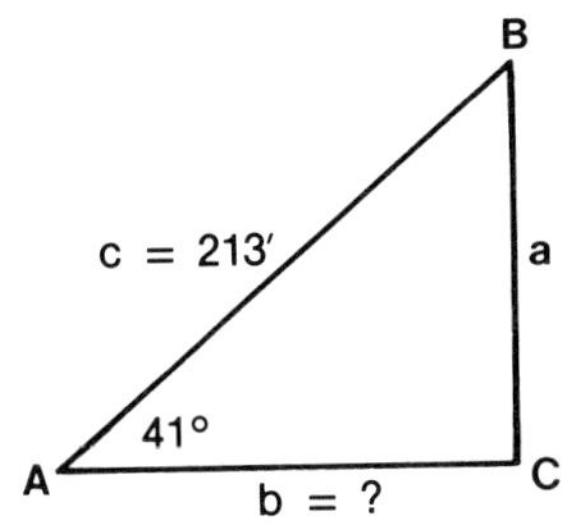

Fig. 7.4

Tangent. The tangent of an acute angle in a right triangle is found by dividing the side opposite the angle by the side adjacent to the angle:

$$\text{tangent} = \frac{\text{opposite side}}{\text{adjacent side}} .$$

Referring to figure 7.2, the tangent of angle A (tan A) is the ratio of the side opposite angle A to the side adjacent to angle A, or

$$\tan A = a/b.$$

Likewise, the tangent of angle B is the ratio of the side opposite angle B and the side adjacent to angle B, or

$$\tan B = b/a.$$

The tangent of an angle changes when the angle changes. As the angle increases, the tangent increases. If the angle is very small, the side opposite the angle is very short and thus the tangent of a small angle is small (fig. 7.5). When the angle (B in fig. 7.5) comes close to 90°, the hypotenuse (c) is nearly parallel to the side opposite the angle (b).

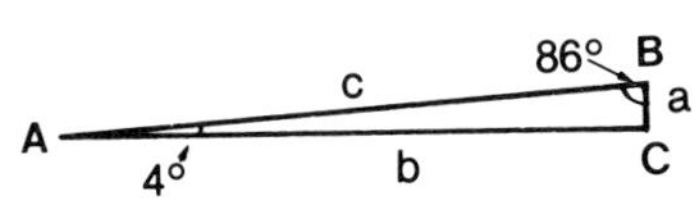

Fig. 7.5. The ratio of a to b is 0.0699.

The tangent function is often used to find the angle of elevation and the angle of depression. In figure 7.6, angle A is the angle of elevation. The observer's line of sight is from A to B. When the observation is made from a high point (A in figure 7.7) down to a low point (B), the angle at A is called the angle of depression.

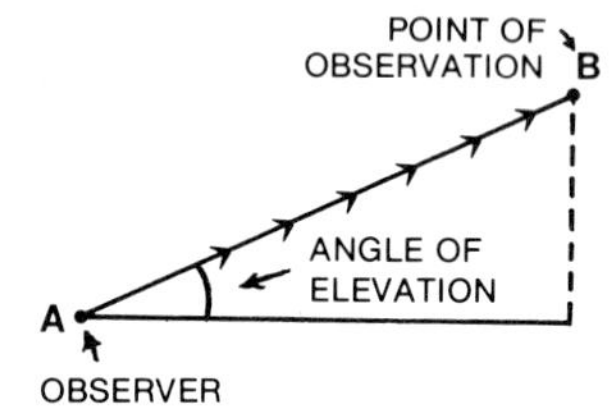

Fig. 7.6. Angle of elevation

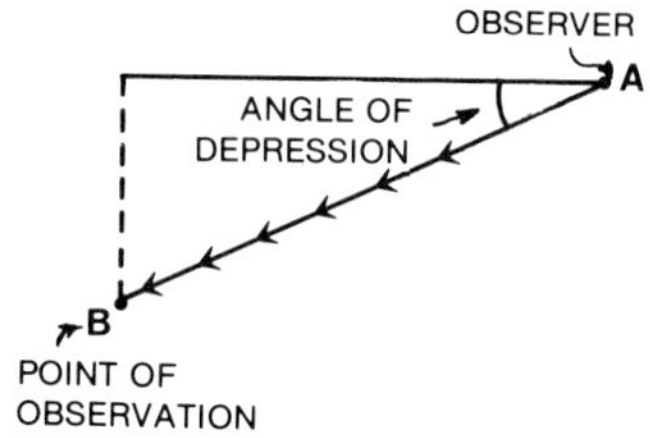

Fig. 7.7. Angle of depression

Practice Problem: Find the angle of elevation (to the nearest degree) of a ramp built from the ground up to a building doorway. The ramp rises 1.5 feet in a horizontal distance of 8 feet, as shown in figure 7.8.

Solution: Use the tangent function and substitute known values:

$\tan A = a/b$

$\tan A = 1.5' \div 8' = 0.1875.$

Using table 7.1, the nearest number to 0.1875 under the "Tangent" column is 0.1944, which is the tangent for 11°. The ramp rises at an 11° angle from the ground, or its angle of elevation is 11°.

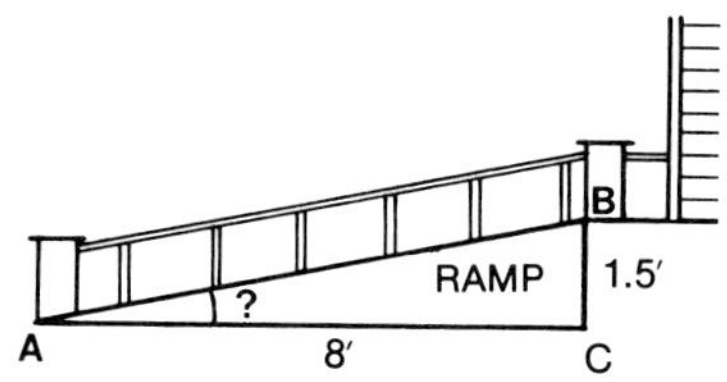

Fig. 7.8

Reciprocals. Three other trigonometric functions, known as the cosecant (csc), secant (sec), and cotangent (cot), are the reciprocals of the sine, cosine, and tangent, respectively. The reciprocal of any number is 1 divided by that number; for example, the reciprocal of 20 is 1/20, or 0.05. In terms of figure 7.2, the reciprocals are as follows:

$$\text{cosecant} = \frac{\text{hypotenuse}}{\text{opposite side}}$$

$$\text{secant} = \frac{\text{hypotenuse}}{\text{adjacent side}}$$

$$\text{cotangent} = \frac{\text{adjacent side}}{\text{opposite side}} .$$

How these compare with the three main functions is shown in table 7.2.

Table 7.2
Reciprocal Functions

Function	Reciprocal		
$\sin A = a/c$	$\csc A = 1/\sin A$	or	$\csc A = c/a$
$\cos A = b/c$	$\sec A = 1/\cos A$	or	$\sec A = c/b$
$\tan A = a/b$	$\cot A = 1/\tan A$	or	$\cot A = b/a$
$\sin B = b/c$	$\csc B = 1/\sin B$	or	$\csc B = c/b$
$\cos B = a/c$	$\sec B = 1/\cos B$	or	$\sec B = c/a$
$\tan B = b/a$	$\cot B = 1/\tan B$	or	$\cot B = a/b$

It is interesting to note that in a right triangle the sine of one acute angle is the cosine of the other, and that the tangent of one angle is the reciprocal of the tangent (cotangent) of the other. For example, in table 7.1, the tangent of 30° is 0.5774. The other acute angle in a right triangle would be 60°, and the tangent of 60° is 1.7321, which is equal to 1/0.5774, or the reciprocal of 0.5774. A scientific calculator can quickly find the reciprocal functions of an angle. This function key is usually designated as $1/x$. Other relations among the six functions of an angle exist and play an important role in areas of trigonometry that will not be covered in this limited study.

Inverse functions. When a function of an angle is known but the angle is not, the inverse function is used to find the value of the angle. The inverse function *represents an angle.* Thus, an angle whose function is x is the inverse function of the angle. For example, the angle whose sine is 0.7071 is 45° (see table 7.1). The symbol for the expression "an angle whose sine is x" is arc sin x or $\sin^{-1}x$. On a scientific calculator the key to find the inverse function may be designated with one of these symbols, or it may simply be labeled "INV." To use this function, the number is entered, the inverse key is pressed, and then the desired function key (SIN, COS, TAN) is entered. On the calculator, the angle whose sine is 0.7071 is shown to be 44.999451.

Example Problem: Find the angle whose tangent is 2.1546.

Solution: Using table 7.1, find the nearest numbers in the "Tangent" column to 2.1546. These numbers are 2.1445 (65°) and 2.2460 (66°). Interpolate to find the exact angle.

$$2.1546 - 2.1445 = 101$$
$$2.2460 - 2.1546 = 914$$
$$101 \div 914 = 0.11.$$

Since 2.1445 is closer to 2.1546 than 2.2460, the angle will be

$$65° + 0.11 = 65.11°.$$

The calculator will find the answer to be 65.1028°.

APPLYING TRIGONOMETRIC FUNCTIONS

The trigonometric ratios are of little value unless they are applied appropriately. For example, if the hypotenuse and one acute angle are known, the sine is used to find the opposite side, and the cosine is used to find the adjacent side.

Example Problem: In the right triangle in figure 7.9, angle $A = 34°$ and side $c = 150'$. Find the lengths of sides a and b.

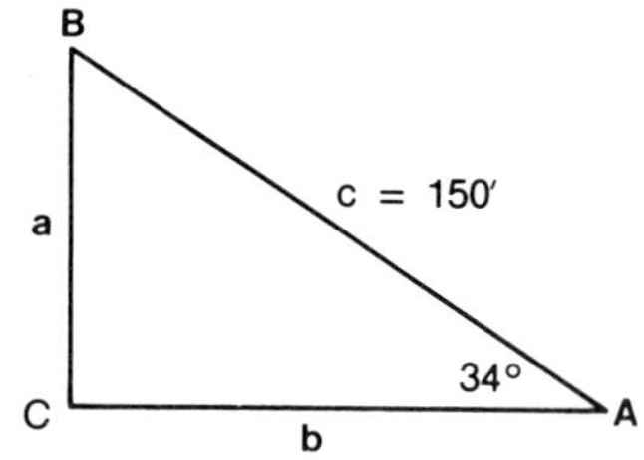

Fig. 7.9

Solution: To find side a, use

$\sin A = a/c.$

Transpose to

$a = c \times \sin A.$

From table 7.1, the sine of 34° is 0.5592.

Substitute numerical values:

$a = 150 \times 0.5592 = 83.88'.$

To find side b, use

$\cos A = b/c$

$b = c \times \cos A$

From table 7.2, the cosine of 34° is found to be 0.8290.

$b = 150 \times 0.8290 = 124.35'.$

If an acute angle and the opposite side are known, the hypotenuse can be found by using the sine, and the adjacent side can be found by using the tangent.

Example Problem: In the right triangle in figure 7.10, angle $A = 40°$ and side $a = 37''$. Find sides c and b.

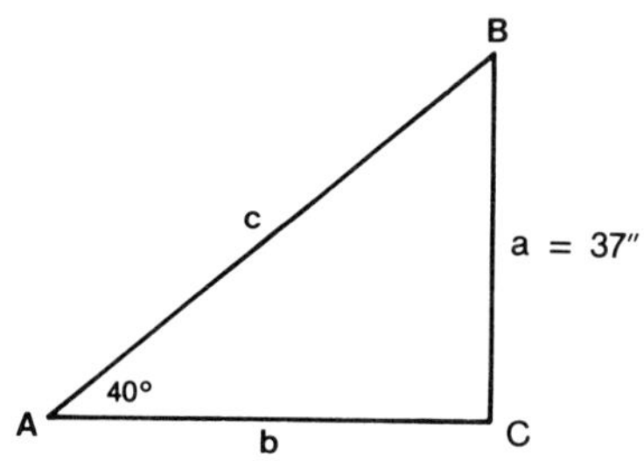

Fig. 7.10

Solution: To find side c:

$\sin A = a/c$

$\sin 40° = 0.6428$

$c = a \div \sin A = 37 \div 0.6428 = 57.56''.$

To find side b:

$\tan A = a/b$

$\tan 40° = 0.8391$

$b = a \div \tan A = 37 \div 0.8391 = 44.1''.$

These functions may be used in many other combinations to solve for missing measurements. To choose the correct function for any problem, find the function that contains *two known quantities* and the *unknown quantity desired.*

The size of an angle may be found if two of the sides of the triangle are known. Using the values of the sides, the value of the trigonometric function of the angle can be found. Then, using the inverse function, the angle that matches the value of the function can be found in the table of functions (table 7.1) or by using a scientific calculator.

Example Problem: In a right triangle, the two legs are 13 inches and 9 inches long. Find the smaller acute angle.

Solution: Drawing such a triangle (fig. 7.11) will show that the smaller angle is opposite the shorter side. Then the tangent of the unknown angle is 9/13, or 0.6923. Using the inverse function find an angle whose tangent is 0.6745 (arc tan 0.6745, or $\tan^{-1} 0.6745$). Using the calculator, the angle is found to be 33.9996. The table of functions shows that tan 34° = 0.6745 and tan 35° = 0.7002. Since 0.6923 is slightly closer to 0.7002 than it is to 0.6745, the unknown angle is found to be about 34.6°, or 34°36′.

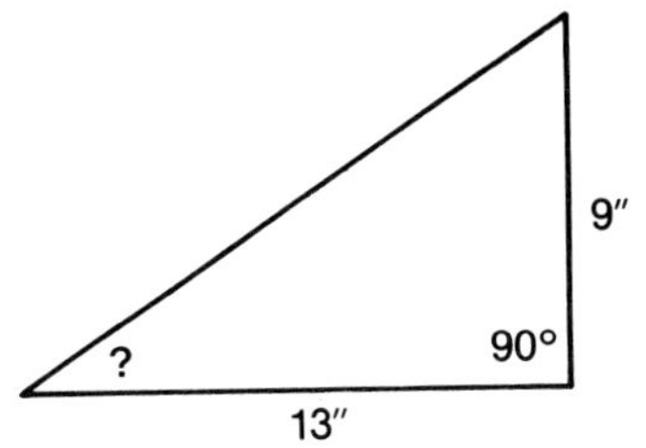

Fig. 7.11

PRACTICE PROBLEMS

Round off answers to two places.

1. A right triangle has one 15° angle. How many degrees does the other acute angle contain? ____________

2. Find the length of side c in the triangle shown below. ____________

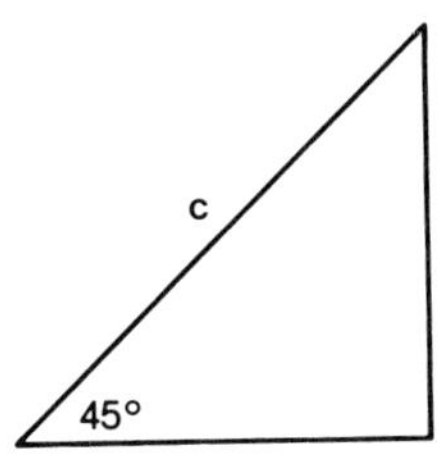

3. The guy wire to the top of a gin pole that is 30 feet high makes an angle of 45° with the ground. What is the length of the wire? ____________

4. Find the length of the hypotenuse in the accompanying sketch. ____________

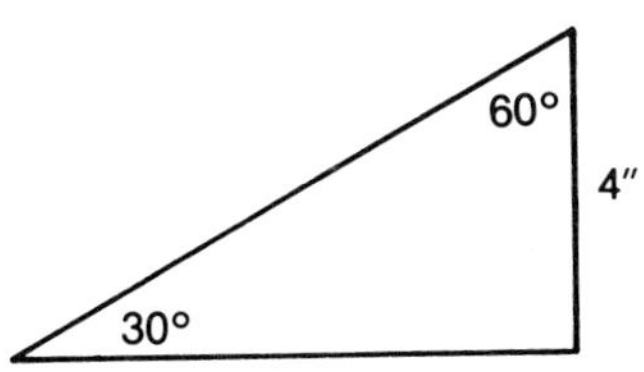

5. Find the length of the side opposite the 60° angle in the sketch in problem 4. ____________

6. The side opposite a 36° angle in a right triangle is 12 inches long. How long is the hypotenuse? ____________

7. Find the length of the pieces of galvanized roofing needed to cover the roof in the sketch below. (Note: Treat as two right triangles.) ______________

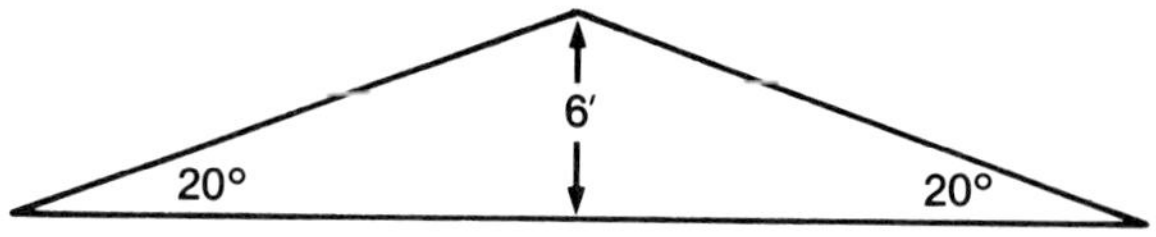

8. A 30-foot ladder is placed against a house at an angle of 48° to the ground.

   a. How high is the top of the ladder above the ground? ______________

   b. How far is the bottom of the ladder from the house? ______________

9. A guy wire to a derrick is 200 feet long and is anchored to the ground 150 feet from the derrick. What angle does the wire make with the ground? ______________

10. In the wedge shown below, find the value of angle *A* between the two sides. ______________

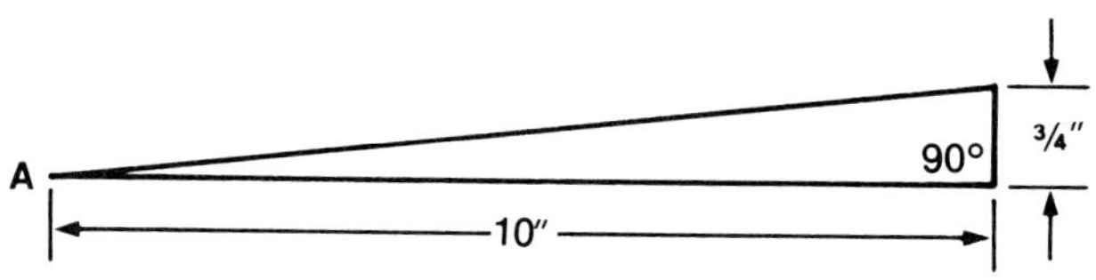

Oblique Triangle Calculations

Right triangle calculations in trigonometry will solve most of the problems encountered in petroleum or other industrial work. Usually, oblique triangles, as well as other shapes, can be divided into right triangles, and the problems can be solved using the trigonometric functions given. A good approach to everyday situations is to find the triangle or triangles in the application and then solve for missing sides and angles.

Example Problem: Determine the largest square that can be milled from a 6-inch circular disk.

Solution: To help visualize the problem, draw a circle of any size and inscribe a square inside it (fig. 7.12). Then, passing through the center of the circle, connect two opposite corners of the square,

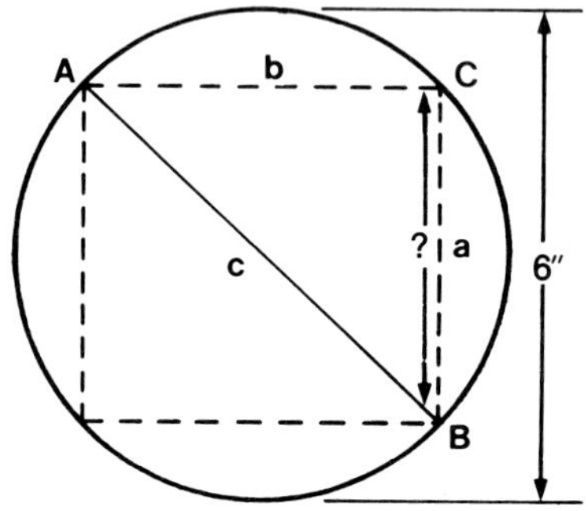

Fig. 7.12

making a right triangle. Label the right angle C and the other angles and sides as shown in figure 7.12. Determine known dimensions: side $c = 6''$; angle $C = 90°$; angles A and $B = 45°$ (half of a right angle). Now, use the sine function,

$$\sin A = a/c,$$

and transpose to solve for a:

$$a = \sin A \times c$$
$$a = \sin 45° \times 6'' = 0.7071 \times 6 = 4.24''.$$

Thus, the largest square that can be milled from a 6-inch circular disk has 4.24-inch sides.

SOLVING FOR SIDES AND ANGLES

Unknown measurements of oblique triangles that have two sides and one angle, two angles and one side, or all three sides known can be found by using the following trigonometric formulas. Reference points for the formulas are shown in figure 7.13.

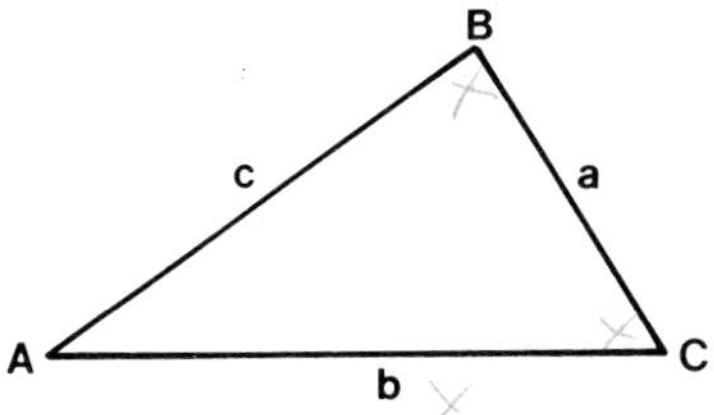

Fig. 7.13

When two angles and one side are known, the other angle and the other two sides can be found. Assuming the known side is b and the known angles are B and C, the formulas are

$$A = 180° - (B + C)$$
$$c = \frac{b \sin C}{\sin B}$$
$$a = \frac{b \sin A}{\sin B} .$$

Example Problem: Find angle B and sides b and c in the triangle shown in figure 7.14.

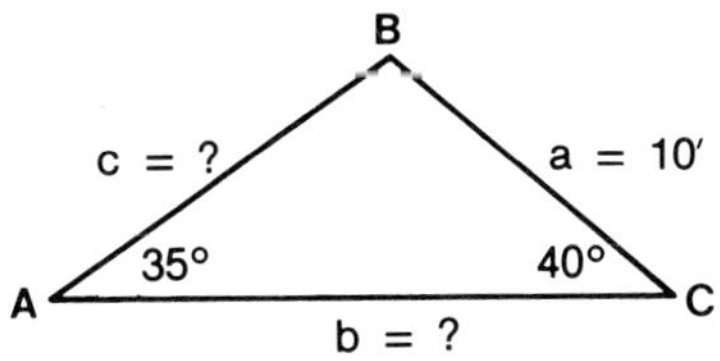

Fig. 7.14

Solution: Transpose the formula to solve for B:

$$B = 180° - (A + C)$$
$$B = 180° - (35° + 40°) = 105°.$$

Substitute the known side a in the formulas given:

$$c = \frac{a \sin C}{\sin A}$$
$$c = (10' \times \sin 40°) \div \sin 35°$$
$$c = (10' \times 0.6428) \div 0.5736 = 11.2064, \text{ or } 11.21'.$$

To solve for side b, it becomes necessary to use the sine of angle B. Angle B is over 90° and therefore an obtuse angle. To find the sine of an obtuse angle, subtract the value of the obtuse angle (in this case,

105°) from 180° and use the sine of the difference (75°) in the equation:

$$b = \frac{a \sin B}{\sin A}$$

$$b = (10' \times \sin 75°) \div \sin 35°$$

$$b = (10' \times 0.9659) \div 0.5736 = 16.8392, \text{ or } 16.84'$$

Thus, angle $C = 40°$; side $c = 11.21'$; and side $b = 16.84'$.

When two sides and the included angle are known, the other two angles and side can be found. Assume the known sides are b and c and the known angle is A. Then

$$a^2 = b^2 + c^2 - 2bc \cos A.$$

$$\sin B = \frac{b \sin A}{a}.$$

$$C = 180° - (A + B).$$

When two sides and the angle opposite one of these sides is known, the other two angles and side can be found. Assume the known angle is B and the known sides are b and c. Side b is opposite angle B. The formulas are

$$\sin C = \frac{c \sin B}{b}.$$

$$A = 180° - (B + C).$$

$$a = \frac{b \sin A}{\sin B}.$$

When all three sides are known, the angles can be found. With sides a, b, and c known, angles A, B, and C can be found using these formulas:

$$\cos B = \frac{c^2 + a^2 - b^2}{2ca}.$$

$$\sin C = \frac{c \sin B}{b}.$$

$$A = 180° - (B + C).$$

FINDING AREAS

When two sides and the included angle of a triangle are given, the area of that triangle can be computed by using the sine of the angle. Assuming sides b and c and angle A (fig. 7.15) are known, the formula for finding the area is

$$\tfrac{1}{2}bc \sin A$$

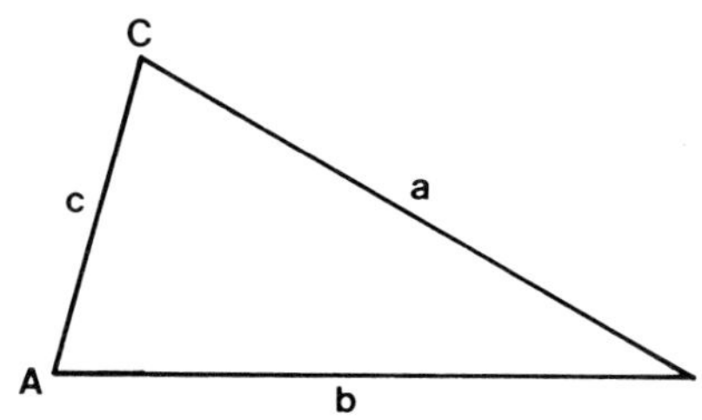

Fig. 7.15

This formula applies if the angle involved is an acute angle. If the included angle is obtuse, the formula becomes

$$\text{area} = \tfrac{1}{2}ac \sin (180° - B)$$

or

$$\text{area} = \tfrac{1}{2}bc \sin (180° - A).$$

PRACTICE PROBLEMS

Round off answers to two places.

1. The equal sides of a roof make an angle of 35° with the horizontal. If each side of the roof is 50 feet, how long is the span of the roof? __________

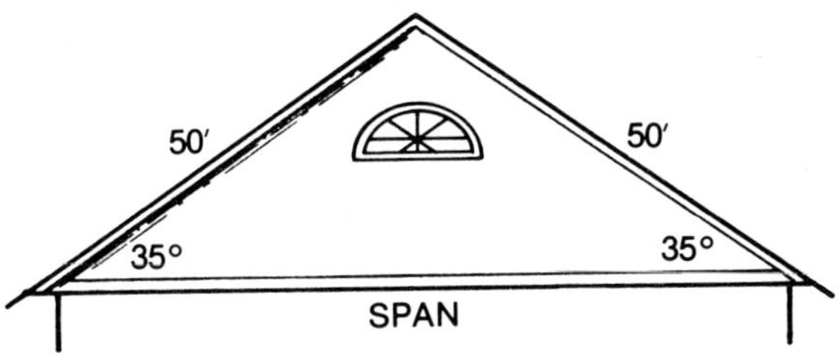

2. Find the length of side a in the triangle shown below. __________

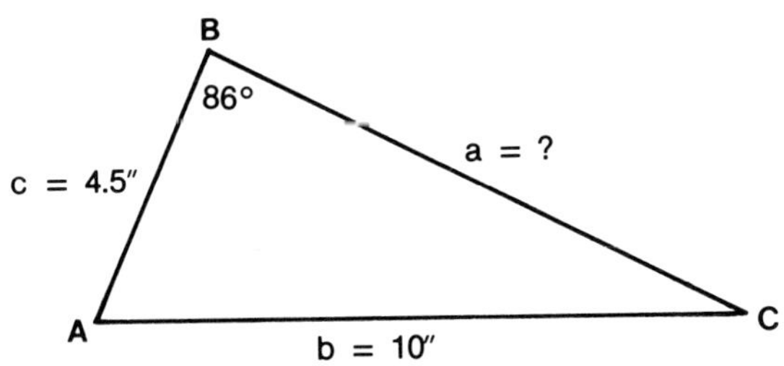

3. Find the angle whose cosine is 0.5240. __________

4. An oblique triangle has a 4″ side and a 6″ side that join to make a 65° angle. What is the measurement of the other side? __________

5. Using the sine and cosine, find the dimensions of X and Y in the sketch below. ______ ______

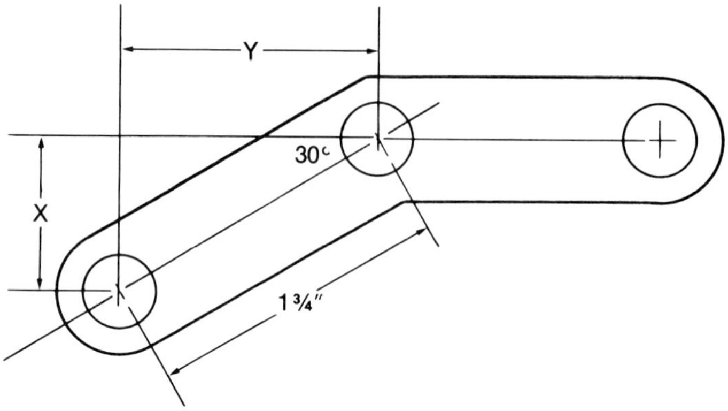

6. Using the sine, find dimension c in the plate sketched below. ______

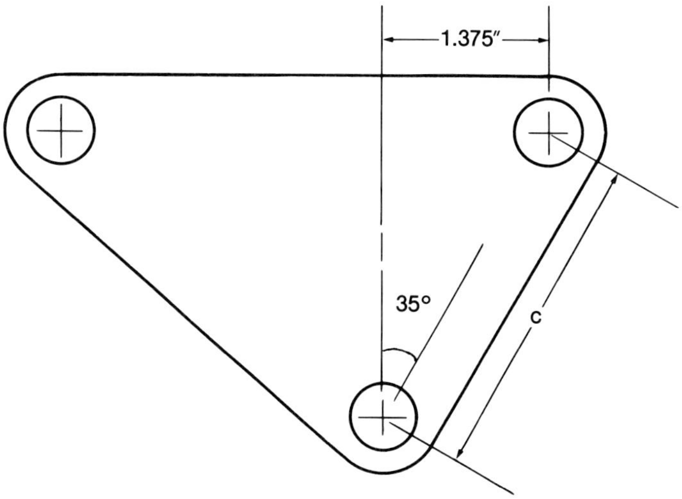

7. Find the area of the triangle below, using trigonometric functions. ______

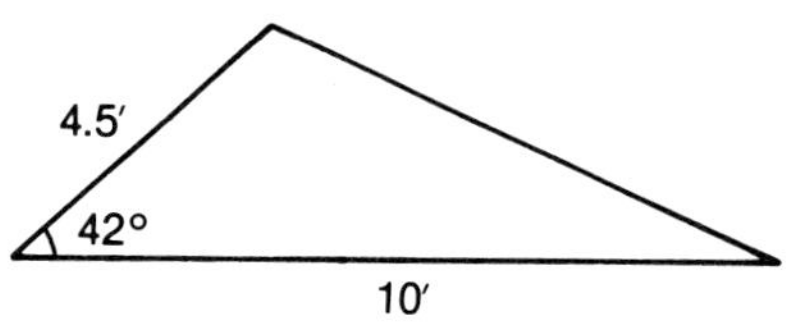

8. Approximately how many square feet are in a triangular plot of ground that is 20′ × 30′ × 35′? ____________

9. The derrick legs in the sketch below make an angle of 10° with the vertical. Calculate their length. ____________

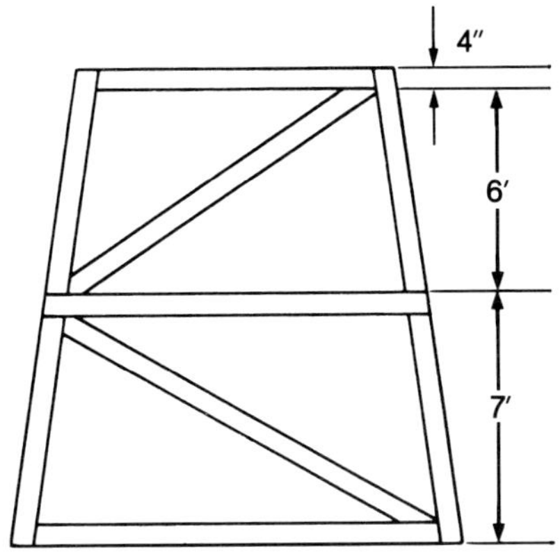

10. What is the length of each side of the largest pentagon that can be cut from a circular disk that is 6 inches in diameter? ____________

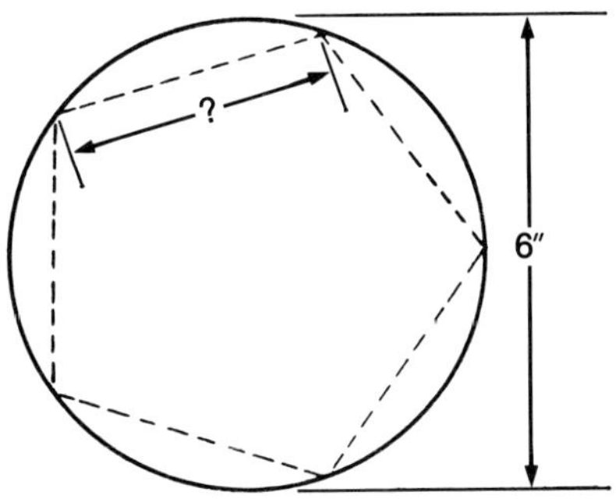

SELF-TEST
7. Trigonometry

Multiply the number of problems solved correctly by five to arrive at your percentage of competency.

Round off answers to one place.

1. What is the length of side *b* in the sketch below?

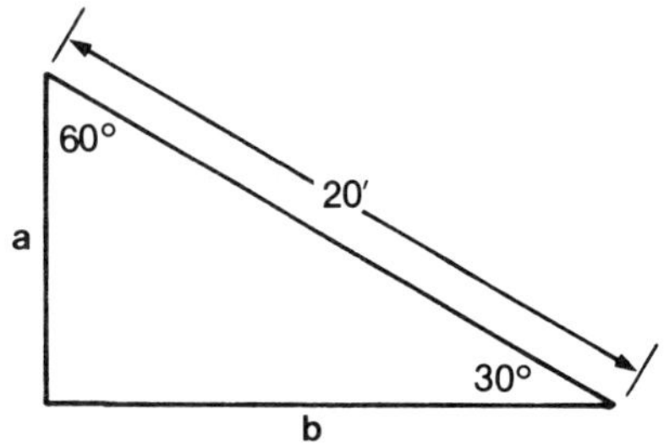

____

2. Find the lengths of the two guy wires shown in the accompanying sketch.

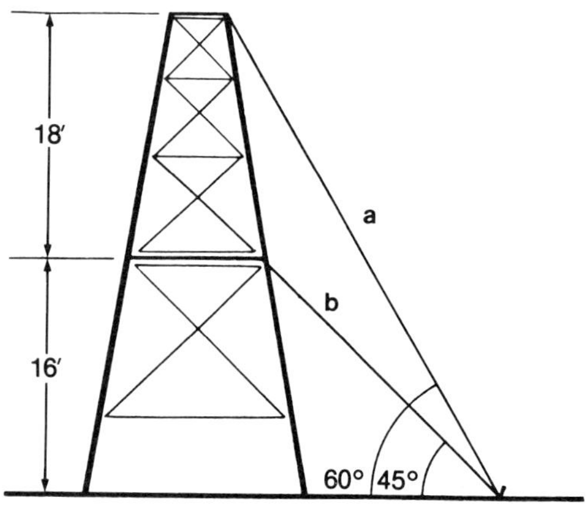

____

3. Find the length of the guy wire in the sketch below.

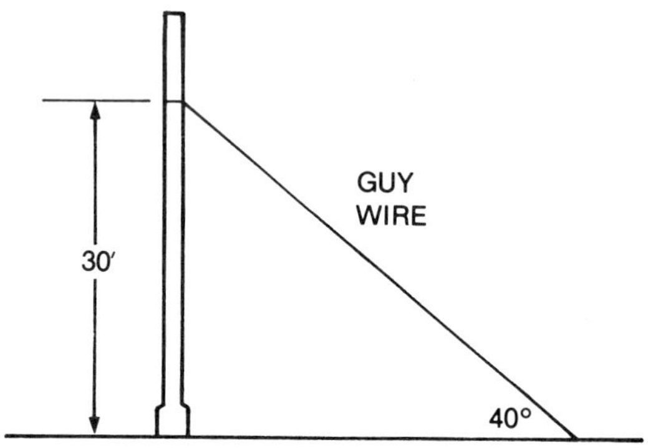

____

4. A right triangle has angles of 28° and 62°. The hypotenuse is 12 inches long. How long is the side opposite the 28° angle?

____

5. Find the length of the hypotenuse of the triangle shown below.

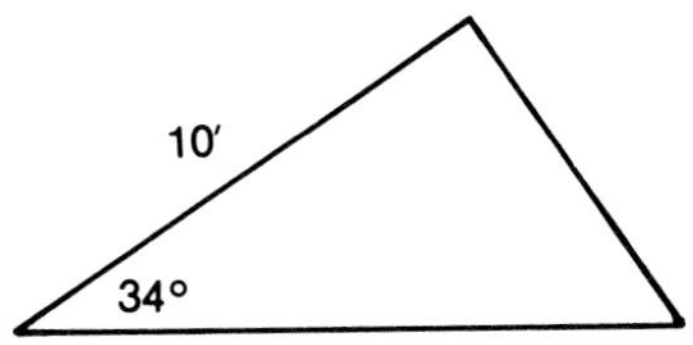

____

6. A balloon that is 600 feet high when straight up is driven by the wind until its tie rope is at an angle of 80° to the ground. What is the height of the balloon at this point?

____

7. Calculate the following distances on the sketch below:

a. Length of the guy wire ______________________________

b. Height from top of the pole to the ground ______________________________

c. Height of the pole ______________________________

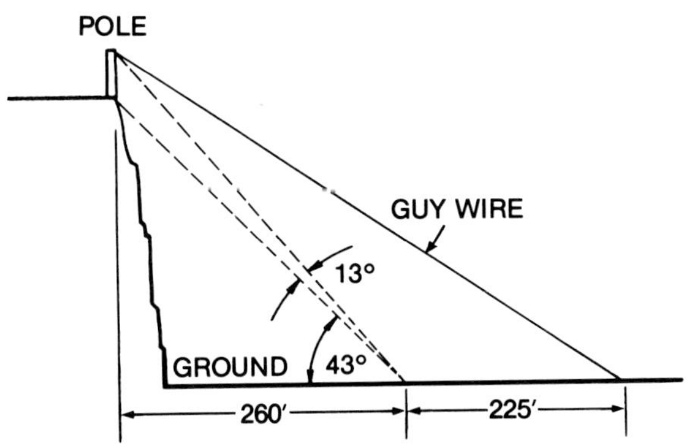

8. Find the angles of the triangle shown in the sketch below.

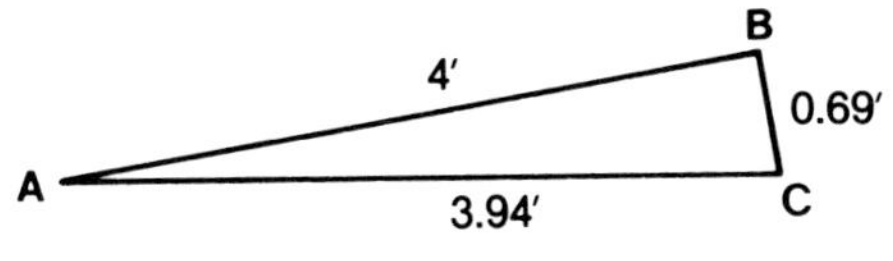

Angle A: ______________________________

Angle B: ______________________________

Angle C: ______________________________

9. Find the length of side a on the sketch below.

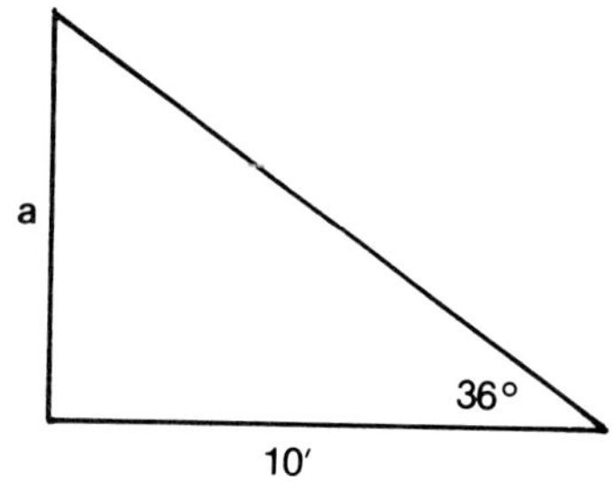

10. If a right triangle has an angle of 46° and the side opposite the angle is 9 feet, what is the length of the adjacent side?

11. Find the length of the center section of the A-frame shown in the sketch.

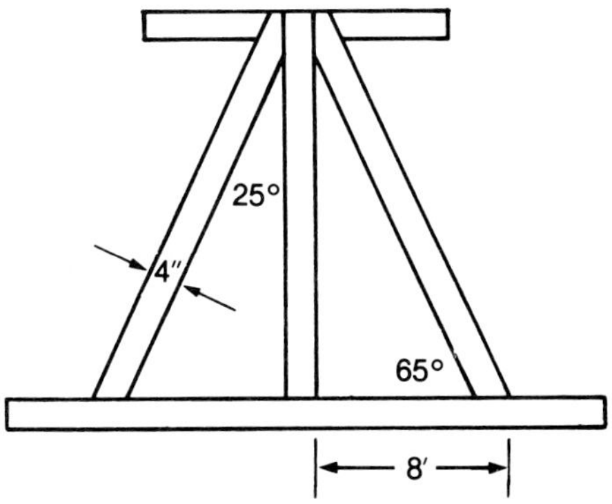

12. The plot of ground shown in the sketch below is cut in two by a river. Side b and the 32° and 90° angles are known, but it is impossible to reach point X, which is marked by a stake across the river. Calculate lengths of sides a and c.

Side a: ______

Side c: ______

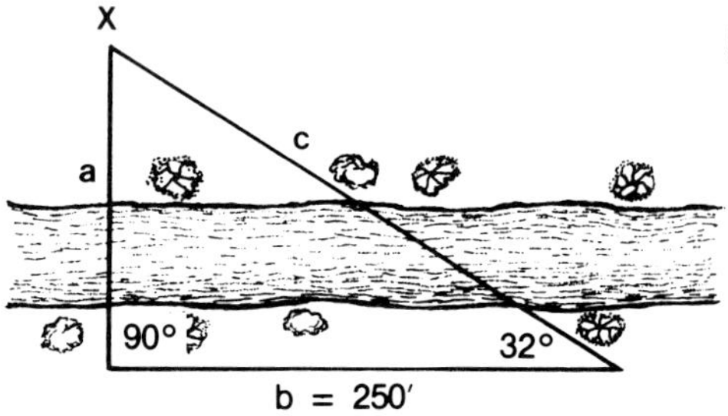

13. Find the distance between the centers for the pulleys shown in the sketch.

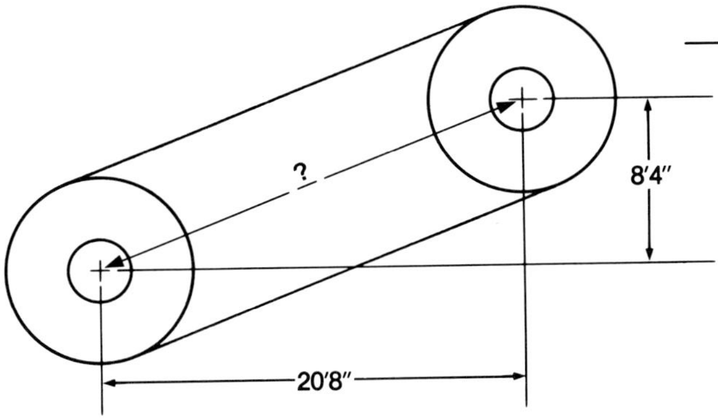

____

14. One of the acute angles in a right triangle is 39°.

a. What is the cotangent of the 39° angle? ______________________

b. What is the cotangent of the other acute angle in the triangle?

____

15. What is the length of each side of the largest hexagon that can be milled from a 6-inch circular disk?

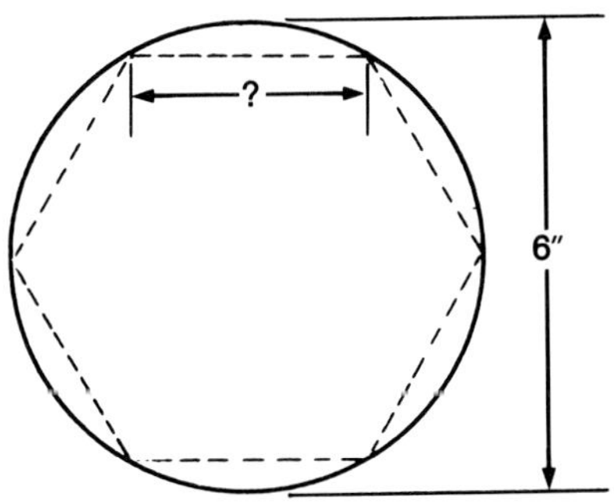

____

16. What is the length of each side of the largest octagon that can be milled from a 6-inch circular disk?

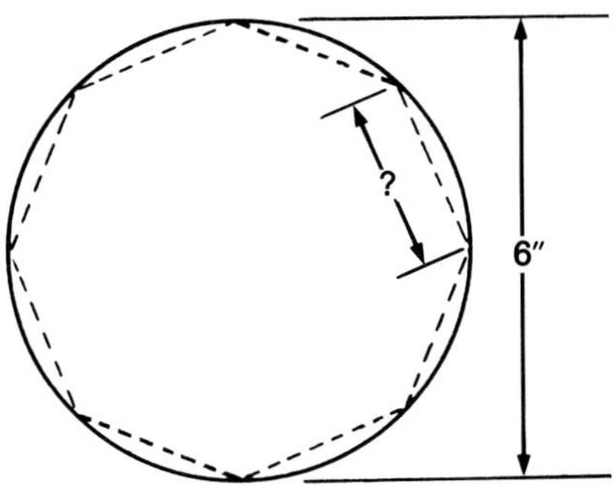

____

17. Find the radius of a circle circumscribed about an octagon with 10-inch sides.

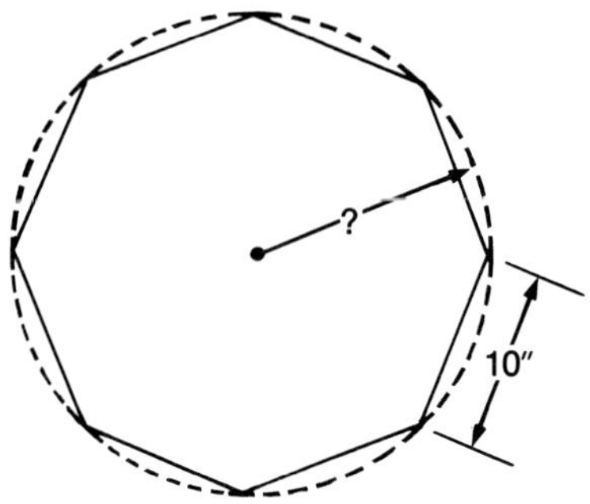

____

18. Nine holes are to be drilled on the circumference of a circle that is 7.25 inches in diameter. What should be the distance between the centers of the holes?

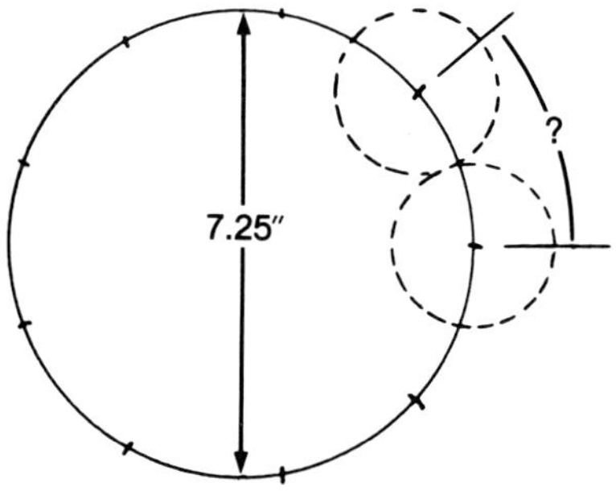

____

19. Find the distance at which to set dividers for laying off twelve equally spaced holes on a 9-inch diameter circle.

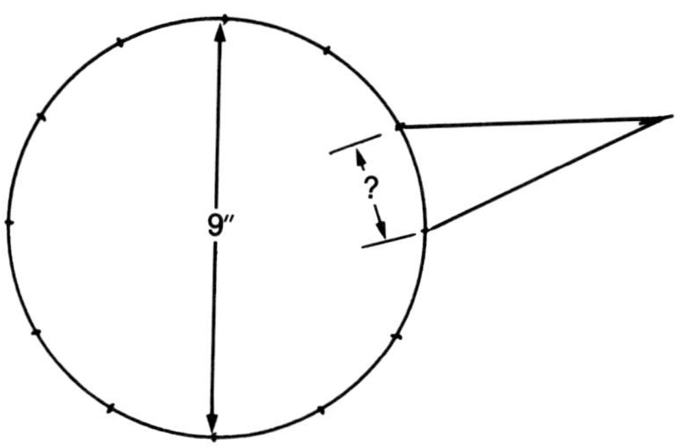

____

20. What is the diameter of a circle circumscribed about a square that is 7 inches on a side?

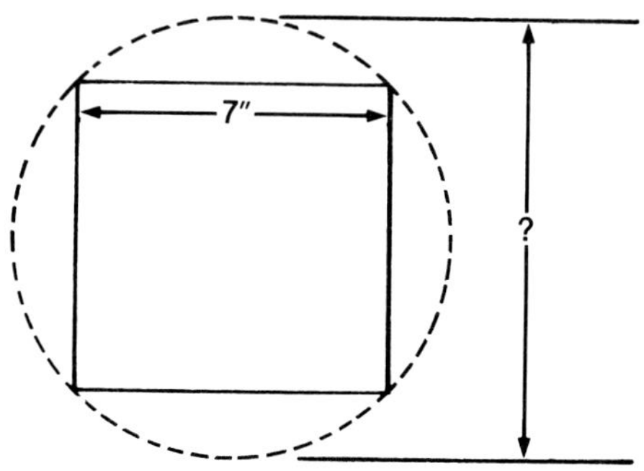

____

8
ADVANCED OIL INDUSTRY APPLICATIONS

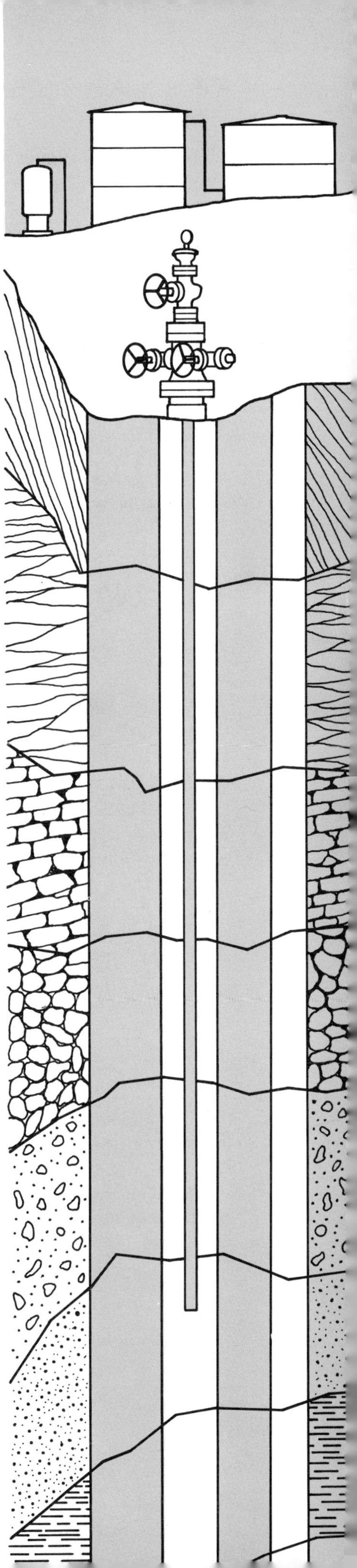

Objectives

Upon completion of chapter eight, the student will be able to—

1. Calculate mud control problems involving mud in the system, mud weighting, cycle time, and annular volume and velocity of mud.
2. Solve well control problems dealing with hydrostatic pressure, circulating pressure, bottomhole pressure, shut-in drill pipe pressure, maximum allowable surface pressure, and gradients of mud and influx.
3. Determine the ton-miles of service required by a drilling line while making a round trip, making hole, coring, and setting casing.
4. Determine the amount of emulsion-treating chemical needed to use on a lease.
5. Calculate the hourly gas flow through an orifice meter.
6. Calculate negative buoyancy involved in pipeline construction under water.
7. Convert hydrostatic head to pressure and vice versa in dealing with pumps.
8. Find the pump horsepower needed for an oil pipeline.
9. Determine the locations of pipeline pumping stations and plot a profile of ground elevations along a pipeline route.
10. Find the amount of heat required to raise the temperature of a volume of liquid.
11. Calculate the amount of product from a particular refining process, using known data on feeds and other products.

The following equations are included here because they apply principles covered in this book to everyday use in the petroleum industry. Other industry-related formulas used in this text are given in Appendix A.

Drilling

Routine operations account for most situations on the drilling rig where mathematics is important. Standard equations are given for everyday situations. Identifying problems and applying the correct equations are essential to good performance with numbers.

MUD CONTROL

One of the main concerns in drilling is the volume of mud in the system, which includes the mud in the pit and the mud in the hole.

In mud conditioning, it is frequently important to know the time required for the mud to make a cycle from the pump suction to the bottom of the hole and back to the pump suction. It is almost always desirable to add weight material or chemicals at such a rate that the mud will make at least one complete cycle during the treatment. Two factors are involved in cycling time during the treatment: (1) the volume of mud in the hole and the active surface volume, and (2) the output rate of the slush pump.

Amount of mud in the system:

$$V = V_p + V_h \qquad \text{(Eq. 1)}$$

where

V = volume of mud in system, in barrels
V_p = volume of mud in pit, in barrels
V_h = volume of mud in hole, in barrels.

Volume of mud in hole:

$$V_h = d^2 \qquad \text{(Eq. 2)}$$

where

V_h = volume of mud in hole, in barrels per 1,000 feet of hole
d = average diameter of hole, in inches.

Volume of mud in pit:

$$V_p = \frac{l \times w \times h}{5.6} \qquad \text{(Eq. 3)}$$

where

V_p = volume of mud in pit, in barrels
l = length of pit, in feet
w = width of pit, in feet
h = height, or depth, of mud in pit, in feet
5.6 = number of cubic feet per barrel.

Example Problem: A hole is 4,000 ft deep and averages 8¾ in. in diameter. The mud pit contains approximately 600 bbl of mud. Find the approximate amount of mud in the system.

Solution: Find mud in the hole, using equation 2:

$V_h = d^2$
$V_h = 8\frac{3}{4}^2 = 76.5$ bbl/1,000 ft
$V_h = 76.5 \times 4 = 306$ bbl of mud in hole.

Then add mud in hole to mud in the pit:

306 + 600 = 906, or approximately 900 bbl of total mud in system.

Mud cycling time:

$$t = V/pS \qquad \text{(Eq. 4)}$$

where

t = cycling time, in minutes
V = volume of mud in the system, in barrels
p = volume of mud per pump stroke, in barrels
S = pumping rate, in strokes per minute.

Weighting mud with barite:

$$B = \frac{1{,}470\,(W_2 - W_1)}{35 - W_2} \qquad \text{(Eq. 5)}$$

where

B = sacks of barite per 100 bbl of mud
W_1 = initial mud weight, in pounds per gallon
W_2 = desired mud weight, in pounds per gallon.

Values of B under various conditions are tabulated in the upper right half of table 8.1.

Table 8.1
Lightening and Weighting Drilling Mud

Initial Mud Weight (ppg)	Desired Mud Weight (ppg)																	
	9.5	10.0	10.5	11.0	11.5	12.0	12.5	13.0	13.5	14.0	14.5	15.0	15.5	16.0	16.5	17.0	17.5	18.0
9	29	59	90	123	156	192	229	268	308	350	395	442	490	542	596	653	714	778
9.5		29	60	92	125	160	196	234	273	315	359	405	452	503	557	612	672	735
10	43		30	61	93	128	164	201	239	280	323	368	414	464	516	571	630	691
10.5	85	30		31	62	96	131	167	205	245	287	331	376	426	479	531	588	648
11	128	60	23		31	64	98	134	171	210	251	294	339	387	437	490	546	605
11.5	171	90	46	19		32	66	101	137	175	215	258	301	348	397	449	504	562
12	214	120	69	37	16		33	67	103	140	179	221	263	310	357	408	462	518
12.5	256	150	92	56	32	14		34	68	105	144	184	226	271	318	367	420	475
13	299	180	115	75	48	27	12		34	70	108	147	188	232	278	327	378	432
13.5	342	210	138	94	63	41	24	11		35	72	111	150	194	238	286	336	389
14	385	240	161	112	76	54	36	21	10		36	74	113	155	199	245	294	345
14.5	427	270	185	131	95	68	48	32	19	9		37	75	116	159	204	252	302
15	470	300	208	150	110	82	60	43	29	18	8		37	77	119	163	210	259
15.5	513	330	231	169	126	95	72	54	39	26	16	8		39	79	122	168	216
16	556	360	254	187	142	109	84	64	48	35	24	15	7		40	81	126	172
16.5	598	390	277	206	158	123	96	75	58	44	32	23	14	7		41	84	129
17	641	420	300	225	174	136	108	86	68	53	40	30	21	13	6		42	86
17.5	684	450	323	244	189	150	120	96	77	62	49	38	28	20	12	6		43
18	726	480	346	262	205	163	132	107	87	71	57	45	35	26	18	12	5	

The **lower left half** of this table shows the number of barrels of water that must be added to 100 bbl of mud to produce desired weight reductions. To use this portion of the table, locate the initial mud weight in the vertical column at the left, then locate the desired mud weight in the upper horizontal row. The number of barrels of water to be added per 100 bbl of mud is read directly across from the initial weight and directly below the desired mud weight. For example, to reduce an 11 ppg mud to a 9.5 ppg mud, 128 bbl of water must be added for every 100 bbl of mud in the system.

The **upper right half** of this table shows the number of sacks of barite that must be added to 100 bbl of mud to produce desired weight increases. To use this portion of the table, locate the initial mud weight in the vertical column to the left, then locate the desired mud weight in the upper horizontal row. The number of sacks of barite to be added per 100 bbl of mud is read directly across from the initial weight and directly below the desired mud weight. For example, to raise an 11 ppg mud to 14.5 ppg, 251 sacks of barite must be added per 100 bbl of mud in the system.

Example Problem: In the system above, suppose a 12-ppg mud is to be weighted with barite to a weight of 14.5 ppg. How many sacks of barite are to be added?

Solution: This problem may be solved by using equation 5 or by using table 8.1. From table 8.1, it is seen that 179 sacks are needed to increase 100 bbl from 12 ppg to 14.5 ppg. Then for a 900-bbl system:

$B = 179 \times 9 = 1{,}611$ sacks of barite to be added.

Lightening mud by adding water:

$$X = \frac{100\,(W_1 - W_2)}{W_2 - 8.33} \qquad \text{(Eq. 6)}$$

where

X = water to be added, in barrels per 100 bbl of mud
W_1 = initial mud weight, in pounds per gallon
W_2 = desired mud weight, in pounds per gallon.

Weighting mud with bentonite:

$$B = \frac{874\,(W_2 - W_1)}{20.8 - W_2} \qquad \text{(Eq. 7)}$$

where

B = sacks of bentonite per 100 bbl of mud
W_1 = initial mud weight, in pounds per gallon
W_2 = desired mud weight, in pounds per gallon.

Volume increase in weighting:

$$V = \frac{100\,(W_2 - W_1)}{N - W_2} \qquad \text{(Eq. 8)}$$

where

V = volume increase, in barrels per 100 bbl of mud
W_1 = initial mud weight, in pounds per gallon
W_2 = final mud weight, in pounds per gallon
N = weight of weighting material, in pounds per gallon (barite: $N = 35.0$; bentonite: $N = 20.8$).

Example Problem: In the system above, what volume increase is caused by the addition of the barite?

Solution: Use equation 8:

$$V = \frac{100\,(W_2 - W_1)}{N - W_2}$$

$W_2 = 14.5$ ppg; $W_1 = 12.0$ ppg; $N = 35$ ppg.

$$V = \frac{100\ (14.5 - 12.0)}{35 - 14.5} = 12.2 \text{ bbl/100 bbl of mud.}$$

$12.2 \times 9 = 110$ bbl = volume increase caused.

Annular volume of mud:

$$V_a = V_h - D_{dp} \qquad \text{(Eq. 9)}$$

where

V_a = annular volume, in barrels per feet
V_h = volume of mud in hole (open hole capacity), in barrels per feet
D_{dp} = capacity of drill pipe and displacement of drill pipe, in barrels per feet

Annular velocity of mud:

$$v_a = PO/V_a \qquad \text{(Eq. 10)}$$

where

v_a = annular velocity, in feet per minute
PO = pump output, in barrels per minute
V_a = annular volume of mud, in barrels per feet.

or if working with a pump output in *gallons per minute:*

$$v_a = PO \times 0.024/V_a.$$

WELL CONTROL

Well control and mud control are closely related, since the pressure from the drilling fluid, or mud, (called *hydrostatic pressure*) counteracts the pressure of the formation fluid. When the formation pressure increases faster than the hydrostatic pressure or the hydrostatic pressure is reduced, a kick occurs. If prompt action is not taken to control the kick or kill the well, a blowout may occur.

Hydrostatic pressure:

$$HP = W \times TVD \times C \qquad \text{(Eq. 11)}$$

where

HP = hydrostatic pressure, in pounds per square inch gauge
W = mud weight, in pounds per gallon
TVD = true vertical depth of hole, in feet
C = a constant (0.052 for ppg; 0.00695 for lb/ft^3; 0.009806 for kg/m^3).

Example Problem: If a well has a true vertical depth of 4,000 ft and a mud weight of 11.2 ppg, what is the hydrostatic pressure?

Solution: Solve for hydrostatic pressure, using equation 11.

$$HP = MW \times TVD \times 0.052$$
$$HP = 11.2 \times 4{,}000 \times 0.052 = 2{,}329.6 \text{ psi.}$$

Initial circulating pressure:

$$ICP = KRP + SIDPP \qquad \text{(Eq. 12)}$$

where

ICP = initial circulating pressure
KRP = kill-rate pressure
$SIDPP$ = shut-in drill pipe pressure.

Example Problem: If the kill-rate pressure is 900 psi and the initial circulating pressure is 1,110 psi, what is the shut-in drill pipe pressure?

Solution: Use equation 12 and transpose to solve for the shut-in drill pipe pressure:

$$SIDPP = ICP - KRP$$
$$SIDPP = 1{,}110 - 900 = 210 \text{ psi.}$$

Final circulating pressure:

$$FCP = KRP \times W_2 \div W_1 \qquad \text{(Eq. 13)}$$

where

FCP = final circulating pressure
KRP = kill-rate pressure
W_2 = new mud weight
W_1 = old mud weight.

Bottomhole pressure:

$$BHP = HP + SIDPP \qquad \text{(Eq. 14)}$$

where

BHP = bottomhole pressure
HP = hydrostatic pressure
$SIDPP$ = shut-in drill pipe pressure.

Maximum allowable surface pressure, or critical surface pressure:

$$MASP = (FG - MG) \times D_{cs} \qquad \text{(Eq. 15)}$$

where

$MASP$ = maximum allowable surface pressure
FG = fracture gradient in psi per foot
MG = mud gradient in psi per foot
D_{cs} = casing seat depth, in feet.

Kill mud weight, or new mud weight:

$$W_2 = W_1 + WI_{SIDPP} + WI_{OP} \qquad \text{(Eq. 16)}$$

where

W_2 = kill mud weight, or new mud weight
W_1 = original mud weight
WI_{SIDPP} = mud weight increase to compensate for shut-in drill pipe pressure
WI_{OP} = mud weight increase required for overbalance pressure.

Mud weight increase:

$$WI_{SIDPP} = SIDPP \div (TVD \times C) \qquad \text{(Eq. 17)}$$

$$WI_{OP} = OP \div TVD \div C \qquad \text{(Eq. 18)}$$

where

WI_{SIDPP} = mud weight increase to compensate for shut-in drill pipe pressure
WI_{OP} = mud weight increase required for overbalance pressure
$SIDPP$ = shut-in drill pipe pressure
TVD = true vertical depth
OP = overbalance pressure
C = constant of 0.052 for ppg.

Mud gradient:

$$MG = MW \times C \qquad \text{(Eq. 19)}$$

where

MG = mud gradient
MW = weight of mud
C = constant of 0.052 when using ft, ppg, psi; 0.00695 when using ft, lb/ft^3, psi; 0.009806 when using m, kg/m^3, kPa.

Gradient of influx:

$$IG = MG - \frac{MASP - SIDPP}{IH} \qquad \text{(Eq. 20)}$$

where

IG = influx gradient, in psi per foot
MG = mud gradient, in psi per foot
$MASP$ = maximum allowable surface pressure, in psi
$SIDPP$ = shut-in drill pipe pressure, in psi
IH = influx height, in feet.

Influx height:

$$IH = pg/V_a \qquad \text{(Eq. 21)}$$

where

IH = influx height
pg = pit gain, in barrels
V_a = annular volume, or capacity, in barrels per foot.

Pressure and pump rate:

$$P_2 = P_1(S_2 \div S_1)^2 \qquad \text{(Eq. 22)}$$

where

P_2 = new pump pressure, in psi
P_1 = old pump pressure, in psi
S_2 = new pump speed, in strokes per minute (spm)
S_1 = old pump speed, in spm.

DRILLING LINE SERVICE

The total service performed by a drilling line includes primarily work done while making hole, making round trips, coring, and setting casing. The methods for calculating the work done in each of these operations follow.

To use figures 8.1 and 8.2, locate the line representing the drill pipe or collars being used and find the point on that line directly above the mud weight being used. Reading across to the left axis from this point will give the effective weight of the drill pipe or collars in the given mud.

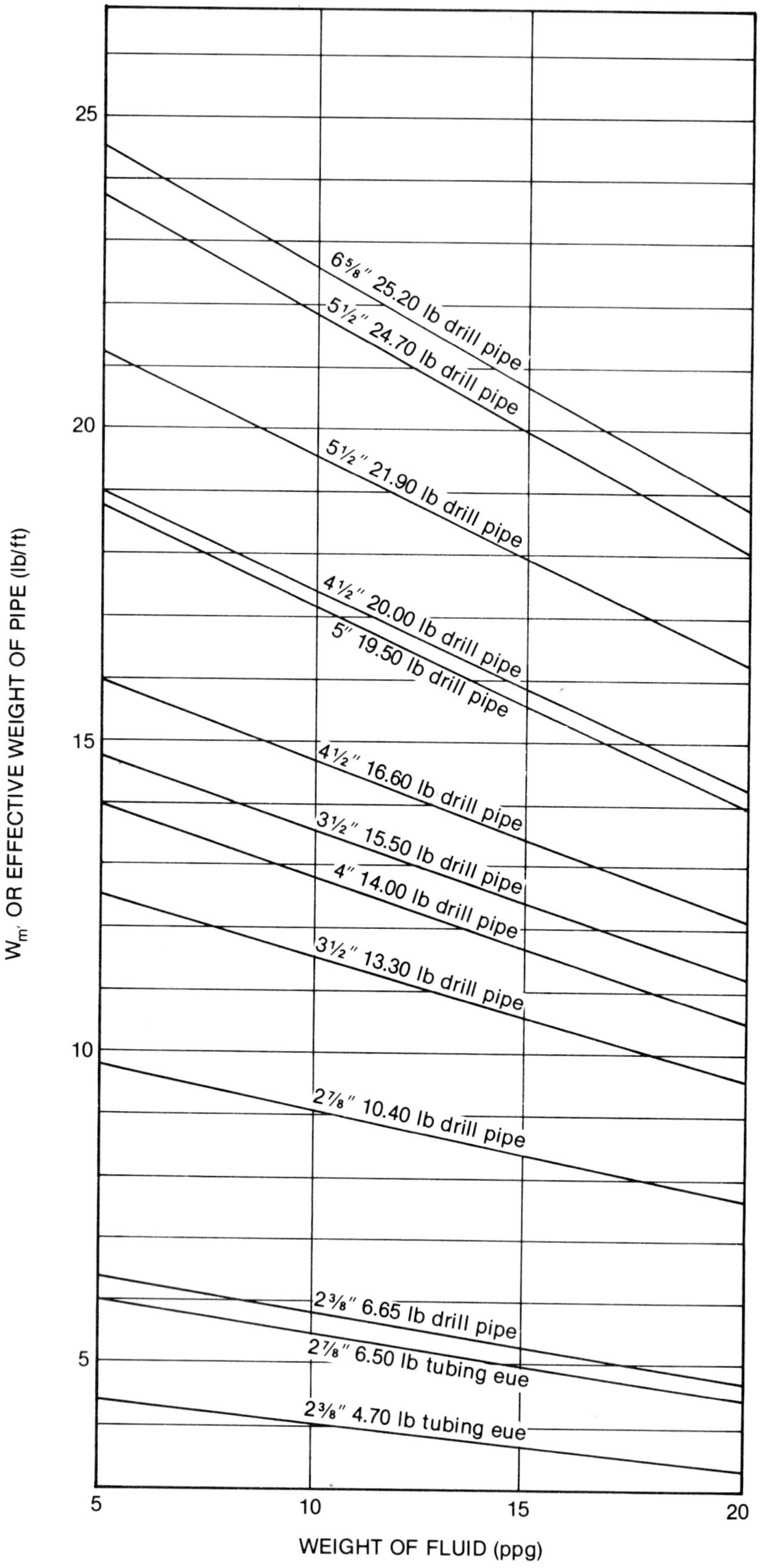

Fig. 8.1. Effective weight of pipe in drilling mud

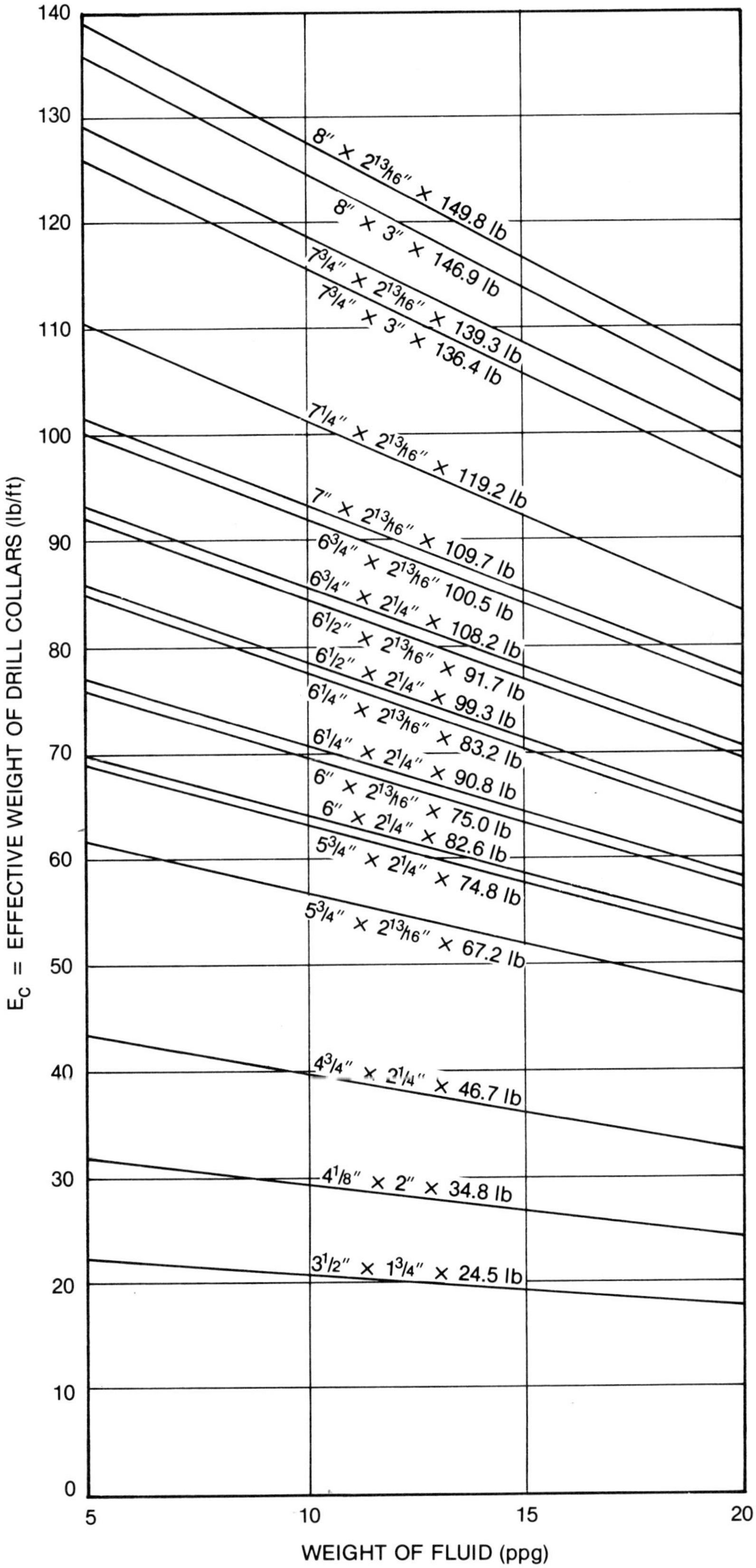

Fig. 8.2. Effective weight of drill collars in drilling mud

Round trip operations:

$$T_r = \frac{W_m D(L_s + D) + 2D(2M + E_c)}{10{,}560{,}000} \qquad \text{(Eq. 23)}$$

where

T_r = ton-miles (weight in tons times distance moved in miles)
D = depth of hole, in feet
L_s = length of drill pipe stand, in feet
W_m = effective weight of drill pipe in mud, in pounds per foot (from fig. 8.1)
M = total weight of traveling block-elevator assembly, in pounds
E_c = effective weight of drill collar assembly in mud (from fig. 8.2) minus the effective weight of the same length of drill pipe in mud (from fig. 8.1).

Example Problem: Find the ton-miles of wear for a round trip made at 5,000 ft with drill collars 200′ × 5¾″ × 2¼″, drill pipe 4½″ – 16.6 lb/ft, M = 10,720 lb, and 10.5-ppg mud.

Solution: Use equation 23:

$$T_r = \frac{W_m D(L_s + D) + 2D(2M + E_c)}{10{,}560{,}000}$$

From figure 8.1:

W_m = 14.5 lb/ft
D = 5,000 ft
L_s = 100 ft (average value)
M = 10,720 lb.

From figure 8.2:

$E_c = (60.7 - 14.5)200 = 9{,}240$ lb.

Then substituting:

$$T_r = \frac{14.5 \times 5{,}000 \times (5{,}000 + 100) + (2 \times 5{,}000) \times (21{,}440 + 9{,}240)}{10{,}560{,}000}$$

= 64 ton-miles.

Drilling operations:

$$T_d = 3(T_2 - T_1) \qquad \text{(Eq. 24)}$$

where

T_d = ton-miles drilling
T_1 = ton-miles for one round trip at depth D_1 (where drilling began)
T_2 = ton-miles for one round trip at depth D_2 (where drilling stopped).

Example Problem: Find the number of ton-miles of service when drilling from 11,000 to 11,500 ft with the following data known:

500 ft of drill collars 5¾" × 2¼"
drill pipe 3½" – 15.5 lb/ft
$M = 24{,}450$ lb
mud weight, 12.2 ppg

Solution: Use equation 24:

$$T_d = 3(T_2 - T_1).$$

Using equation 23, find T_2 (service from a round trip at 11,500 ft) and T_1 (service from a round trip at 11,000 ft):

$$T_2 = 328 \text{ ton-miles}$$
$$T_1 = 304 \text{ ton-miles.}$$

Then substituting:

$$T_d = 3 \times (328 - 304) = 72 \text{ ton-miles.}$$

Coring operations:

$$T_c = 2(T_4 - T_3) \qquad \text{(Eq. 25)}$$

where

T_c = ton-miles coring
T_4 = ton-miles for one round trip at depth D_4 (where coring stopped)
T_3 = ton-miles for one round trip at depth D_3 (where coring began).

Example Problem: Given the same statistics as in the previous problem, find the number of ton-miles of service when coring from 11,000 ft to 11,500 ft instead of drilling.

Solution: Use equation 25:

$$T_c = 2(T_4 - T_3).$$

T_4 in this equation is the same at 11,500 ft as T_2 is in the previous problem, and T_3 at 11,000 ft is identical to T_1 in the previous problem. Thus:

$$T_c = 2 \times (328 - 304) = 48 \text{ ton-miles.}$$

Setting casing operations:

$$T_s = \frac{D(L_{cs} + D)W_{cm} + 4DM}{21{,}120{,}000} \qquad \text{(Eq. 26)}$$

where

T_s = ton-miles setting casing
L_{cs} = length of joint of casing, in ft
W_{cm} = effective weight of casing in mud, in lb/ft (may be estimated from data given on drill pipe in figure 8.1).

Example Problem: Find the number of ton-miles of setting casing if the casing is 6″ − 20 lb/ft at 11,500 ft, M = 24,450 lb, and mud weight is 12.2 ppg.

Solution: Use equation 26 and rearrange:

$$T_s = \frac{DW_{cm}\,(D + L_{cs}) + 4DM}{21{,}120{,}000}.$$

Estimating from figure 8.1,

W_{cm} = 16.7 lb/ft.

L_{cs} averages about 30 ft.

Then

$$T_s = \frac{11{,}500 \times 16.7 \times (11{,}500 + 30) + (4 \times 11{,}500 \times 24{,}450)}{21{,}120{,}000}$$

$$= 158 \text{ ton-miles.}$$

PRACTICE PROBLEMS

1. A well hole is 6,000 ft deep, 8½ in. in diameter, and filled with mud. The mud tanks and surface lines contain 800 bbl of mud. How many 100-lb sacks of barite would be needed to raise the weight of the mud from 11 ppg to 12.5 ppg? __________

2. In the well in problem 1, how much water would have to be added to bring the weight from 12.5 ppg back down to 12 ppg? Use table 8.1 for this calculation. __________

3. A 10,000-ft hole contains 800 ft of 6½" × 2¹³⁄₁₆" drill collars and drill pipe that is 5" O.D., 19.5 lb/ft. The total weight of the traveling block-elevator assembly is 15,000 lb. The length of a stand is 100 ft, and the mud is 15 ppg. Find the ton-miles of wear on the drilling line for a round trip on this well. __________

4. Using the statistics in problem 3, find the ton-miles of wear on the drilling line to drill from 10,000 ft to 12,000 ft. __________

5. Solve for hydrostatic pressure when the mud weight is 13.5 ppg and the true vertical depth is 6,000 ft. __________

6. Find the mud weight when the hydrostatic pressure is 3,245 psi and the true vertical depth is 5,340 ft. __________

7. Find the initial circulating pressure if the kill rate pressure is 1,125 psi and the shut-in drill pipe pressure is 200 psi. __________

8. Find the kill mud weight for a well with the following data:

   Shut-in drill pipe pressure = 310 psi
   True vertical depth = 7,000 ft
   Old mud weight = 9.5 psi
   Overbalance pressure = 200 psi. __________

9. What is the mud gradient for 10-ppg mud? __________

10. A pump has a pressure of 2,500 psi when its speed is 50 strokes per minute. If the speed is reduced to 25 spm, what would be its pressure? __________

Production

PRODUCTIVITY INDEX

The ability of reservoir pressure to move fluids into the wellbore at a given flowing pressure depends on *drawdown*—the differential between shut-in and flowing pressure. Drawdown data is used to plot deliverability for the productivity index or for the inflow performance ratio. The productivity index shows the relationship between pressure drawdown and daily production.

$$PI = PR/(SIBHP - FBHP) \qquad \text{(Eq. 27)}$$

where

PI = productivity index, in barrels per day per psi
PR = production, in barrels per day
$SIBHP$ = initial shut-in bottomhole pressure, in psi
$FBHP$ = flowing bottomhole pressure, in psi.

PUMP HORSEPOWER

Used in oil field gathering systems is the following equation for close approximation of horsepower needed to drive a duplex double-acting pump:

$$hp = (R \times P) \div 48{,}000 \qquad \text{(Eq. 28)}$$

where

hp = horsepower required to drive pump
R = rate of flow, in barrels per day
P = delivery pressure, in psig
48,000 = a constant.

Example Problem: A duplex double-acting pump is designed to deliver 5,000 bbl of oil per day at a pressure of 120 psi. How much horsepower is required to drive the pump?

Solution: Use equation 28 and substitute known values.

$$hp = (5{,}000 \times 120) \div 48{,}000 = 12.5 \text{ hp.}$$

OIL RECOVERY

The formula for finding the amount of oil recoverable from a lease is

$$R = FAtpsr\,(1 - c) \qquad \text{(Eq. 29)}$$

where

R = ultimate recovery in barrels
F = 7,758 barrels per acre-foot
A = acres in the lease
t = average thickness of pay zone, in feet
p = porosity of reservoir rock
s = shrinkage factor
r = recovery factor, depending on type of drive and other reservoir conditions
c = percentage of connate water in pore spaces.

TREATING EMULSIONS

To determine the amount of emulsion-treating chemical to use on a lease, it is necessary to run a series of tests known collectively as a *bottle test.* The major steps in this test are as follows.

1. Prepare a mixture of 3 parts xylol and 1 part methanol to use as a solvent for the chemicals.
2. Choose an emulsion-treating chemical to be tested and add 2 cc of the chemical to 98 cc of the xylol-methanol mixture. The resulting mixture is a 2% solution of the chemical.
3. Put 100 cc of the emulsion in each of three prescription bottles and add 0.2 cc of the 2% solution to one bottle, 0.4 cc to one bottle, and 0.8 cc to one bottle. Shake well, and if heaters are to be used in lease treating, heat the oils in the bottles to the same temperature to be used on the lease. If no bottle shows a clear separation of oil and water, choose another chemical and repeat the procedure. If one bottle clearly shows a better separation than the others, note the amount of solution used in this bottle and use that amount in the following tests.
4. Make up a 2% solution of each of the other chemicals that might be used for treating, and add an amount of solution of each chemical equal to that determined in step 3 to separate 100 cc fresh samples of emulsion. Choose the bottles that show the best results, reduce the amount of each chemical to 0.1 cc less than on the previous test, and repeat the test. Continue this process, reducing the amount of chemical used each time until the one chemical that achieves good separation with the least amount of solution is found.
5. Then use the amount of chemical concentrate needed in the bottle test and the percent concentration of the solution to solve for the

Table 8.2
Converting Bottle Test Results to Chemical Pump Rates per 100 bbl of Emulsion Produced

Compound per 100 cc Emulsion as Determined by Bottle Test (cc)			Concentrated Treating Compound per 100 bbl of Emulsion Produced (pints)
1% Solution	2% Solution	Concentrate	
0.025	0.0125	0.00025	0.084
0.05	0.025	0.0005	0.153
0.1	0.05	0.001	0.336
0.2	0.10	0.002	0.672
0.3	0.15	0.003	1.008
0.4	0.20	0.004	1.344
0.5	0.25	0.005	1.680
0.6	0.30	0.006	2.016
0.7	0.35	0.007	2.352
0.8	0.40	0.008	2.688
0.9	0.45	0.009	3.024
1.0	0.50	0.010	3.360
1.1	0.55	0.011	3.70
1.2	0.60	0.012	4.03
1.3	0.65	0.013	4.27
1.4	0.70	0.014	4.70
1.5	0.75	0.015	5.04
1.6	0.80	0.016	5.38
1.7	0.85	0.017	5.71
1.8	0.90	0.018	6.05
1.9	0.95	0.019	6.38
2.0	1.00	0.020	6.72
2.1	1.05	0.021	7.06
2.2	1.10	0.022	7.39
2.3	1.15	0.023	7.73
2.4	1.20	0.024	8.06
2.5	1.25	0.025	8.40
2.6	1.30	0.026	8.74
2.7	1.35	0.027	9.07
2.8	1.40	0.028	9.41
2.9	1.45	0.029	9.74
3.0	1.50	0.030	10.08
3.1	1.55	0.031	10.42
3.2	1.60	0.032	10.75
3.3	1.65	0.033	11.09
3.4	1.70	0.034	11.42
3.5	1.75	0.035	11.76
3.6	1.80	0.036	12.10
3.7	1.85	0.037	12.43
3.8	1.90	0.038	12.77
3.9	1.95	0.039	13.10
4.0	2.00	0.040	13.44

number of pints needed to treat each 100 barrels of emulsion. The general formula for this relationship is

$$p = 336\,sc \qquad \text{(Eq. 30)}$$

where

p = pints of chemical concentrate (solution) per 100 barrels of emulsion

s = amount of solution needed in bottle test, in cubic centimetres per 100 cubic centimetres of emulsion

c = concentration of solution expressed as a decimal fraction

336 = number of pints per barrel.

Example Problem: Suppose the amount of chemical needed in the bottle test is 0.2 cc per 100 cc of emulsion, and the concentration is 2%. How many pints of treating fluid should you use per 100 bbl of emulsion?

Solution: Use equation 30 and substitute.

$$p = 336\,sc$$

$$p = 336 \times 0.2 \times 0.02 = 1.34.$$

The amount of chemical needed for each 100 barrels of emulsion produced is 1.34 pints. Table 8.2 shows tabulated values of p for various values of s and c.

METER RUNS

To measure gas volumes accurately with an orifice meter, a smooth flow of gas must be provided around the meter. Since bends, ells, and constrictions in pipe are sources of turbulence, straight lengths of pipe must be placed just upstream and downstream from the orifice plate. If space considerations prevent long straight runs of pipe, straightening vanes inserted upstream from the orifice plate will reduce the length of straight pipe necessary. Figure 8.3 and table 8.3 illustrate common sources of turbulence and prescribed minimum lengths of pipe necessary in each case. The boxheads across the top of the table represent the ratio of orifice diameter to nominal pipe diameter. The minimum lengths of straight pipe are stated in terms of nominal pipe diameters.

Example Problem: Two ells not in the same plane are upstream from a 4″ orifice in 10″ pipe. What straight lengths of pipe must be placed upstream and downstream from the orifice? If straightening vanes are used, where should they be located? What lengths of pipe would be necessary with straightening vanes?

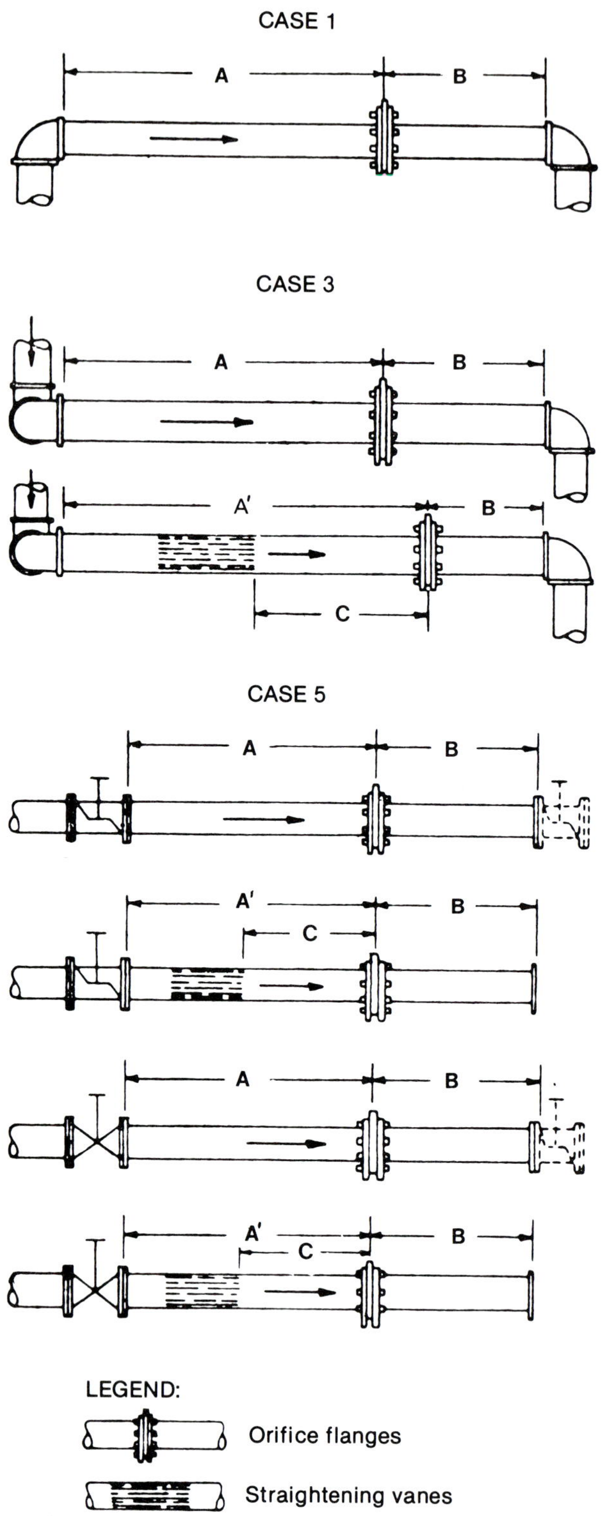

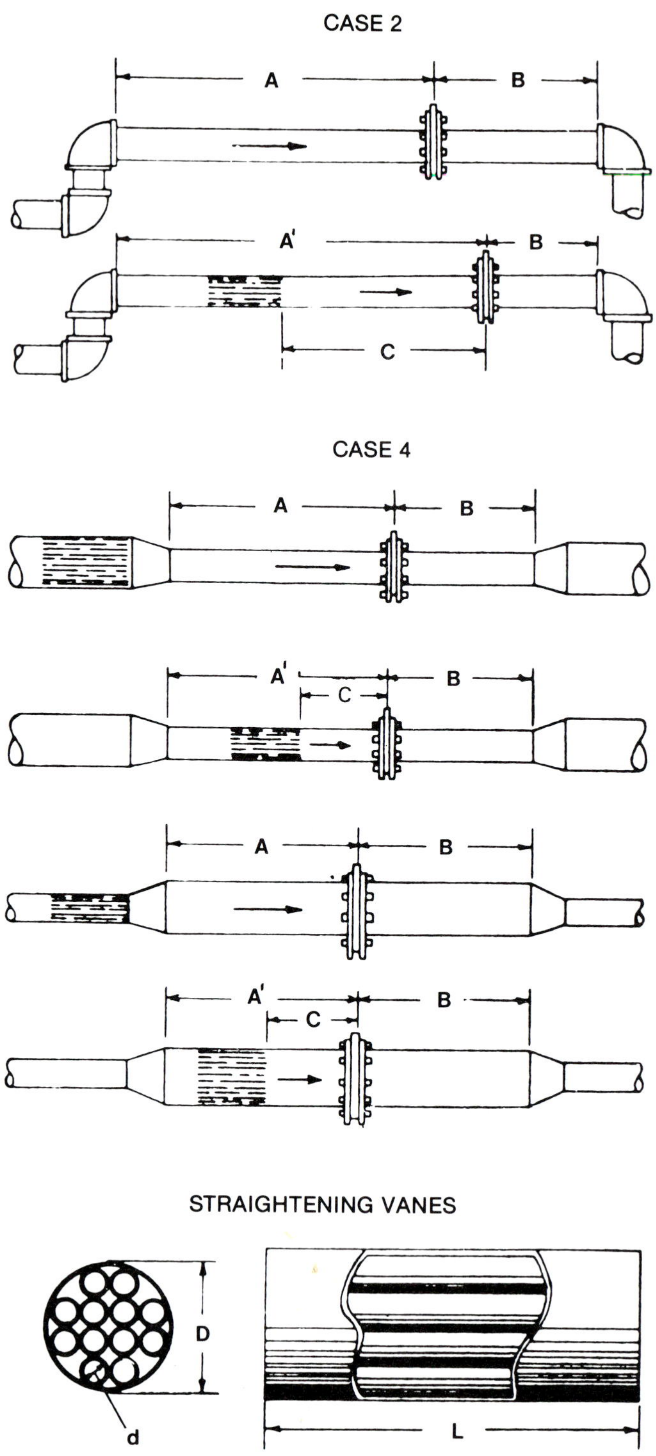

The minimum lengths specified for Case 5 should be used when plug valves are involved.

*The greatest dimension *d* of the inside cross section of any passage through the vanes shall not exceed one-fourth the I.D. of the pipe *D*. The length of the vanes, *L*, shall be at least ten times the dimension *d*.

Fig. 8.3. Standard specifications for meter runs

Solution: The ratio of orifice diameter to nominal pipe diameter is 0.4. The two ells not in the same plane correspond to Case 3 in figure 8.3. From table 8.3, $A = 18.0$ and $B = 3.3$ diameters, or 180″ and 33″, respectively. If straightening vanes are used, B remains the same, but the upstream length is given by $A' = 9.5$ diameters, or 95″. The downstream end of the straightening vanes should be a distance C upstream from the orifice, and from the table, $C = 5.0$ diameters, or 50″.

Table 8.3
Minimum Length Requirements for Meter Runs

		Ratio of Diameters (orifice/pipe)									
	Sec.	0.1	0.2	0.3	0.4	0.5	0.6	0.67	0.70	0.75	0.80
Case 1. One ell or tee ahead of orifice run	A	6.0	6.0	6.0	6.0	6.8	9.0	12.5	14.0	4.4	4.6
	B	2.5	2.8	3.0	3.3	3.7	4.0	4.1	4.3	4.4	4.6
Case 2. Two ells or bends ahead of orifice run (bends in same plane)	A	8.7	8.7	8.7	8.7	10.0	13.5	17.0	19.0	22.0	25.0
	A′	8.7	8.7	8.7	8.7	9.0	10.4	11.6	12.1	14.0	15.5
	B	2.5	2.8	3.0	3.3	3.7	4.0	4.1	4.3	4.4	4.6
	C	5.0	5.0	5.0	5.0	5.1	5.6	6.4	6.7	7.3	8.0
Case 3.* Two ells or bends ahead of orifice run (bends not in same plane)	A	14.0	14.6	16.0	18.0	20.6	25.0	29.0	31.0	35.0	40.0
	A′	9.5	9.5	9.5	9.5	10.0	11.0	12.7	13.5	15.0	16.5
	B	2.5	2.8	3.0	3.3	3.7	4.0	4.1	4.3	4.4	4.6
	C	5.0	5.0	5.0	5.0	5.1	5.6	6.4	6.7	7.3	8.0
Case 4. Reducer ahead of orifice run	A	6.0	6.0	6.0	6.2	7.5	9.6	11.3	12.3	13.7	15.0
	A′	8.7	8.7	8.7	8.7	9.0	10.4	11.8	12.3	13.7	15.0
	B	2.5	2.8	3.0	3.3	3.7	4.0	4.1	4.3	4.4	4.6
	C	5.0	5.0	5.0	5.0	5.1	5.6	6.4	6.7	7.3	8.0
Case 5. Valve or regulator ahead of orifice run	A	16.0	17.8	19.5	21.7	25.0	30.0	35.7	39.0	44.5	50.0
	A′	9.3	9.3	9.3	9.5	10.3	12.0	14.6	15.4	17.0	19.5
	B	2.5	2.8	3.0	3.3	3.7	4.0	4.1	4.3	4.4	4.6
	C	5.0	5.0	5.0	5.0	5.1	5.6	6.4	6.7	7.3	8.0

Note 1. Dimensions given refer to meter-run installations shown in figure 8.3.
Note 2. When pipe taps are used, A, A′, and C should be increased by two pipe diameters and B increased by eight diameters.
Note 3. When the diameter of the orifice may require changing to meet different conditions, the lengths of straight pipe should be those required for the maximum orifice-to-pipe-diameter ratio that may be used.
*Note 4. When in Case 3 the two ells are closely preceded by a third that is not in the same plane as the middle or second ell, the length of A should be doubled.

GAS FLOW

To determine the hourly gas flow through an orifice meter, use the formula

$$Q_h = C' \sqrt{h_w P_f} \qquad \text{(Eq. 31)}$$

where

Q_h = the volume of gas per hour passing the meter, in cubic feet

C' = the orifice flow constant, calculated from several factors, usually stamped on the meter

h_w = differential pressure read from the meter chart, in inches of water

P_f = static pressure, in pounds per square inch absolute.

The static pressure is found by adding the atmospheric pressure at the meter location to the static pressure reading from the meter chart.

Example Problem: The chart in figure 8.4 is from a meter with an orifice flow constant of 800. Find the volume of gas passing the meter between 5 A.M. and 6 A.M. if the atmospheric pressure is 14.5 psi.

Solution: Reading the chart shows the average differential pressure to be about 33.5 in. of water from 5 A.M. to 6 A.M. The static pressure averages about 32 psig during this period, or

$$32 + 14.5 = 46.5 \text{ psia.}$$

Then, using equation 31, the volume of gas is

$$Q_h = 800 \sqrt{33.5 \times 46.5} = 31{,}500 \text{ ft}^3.$$

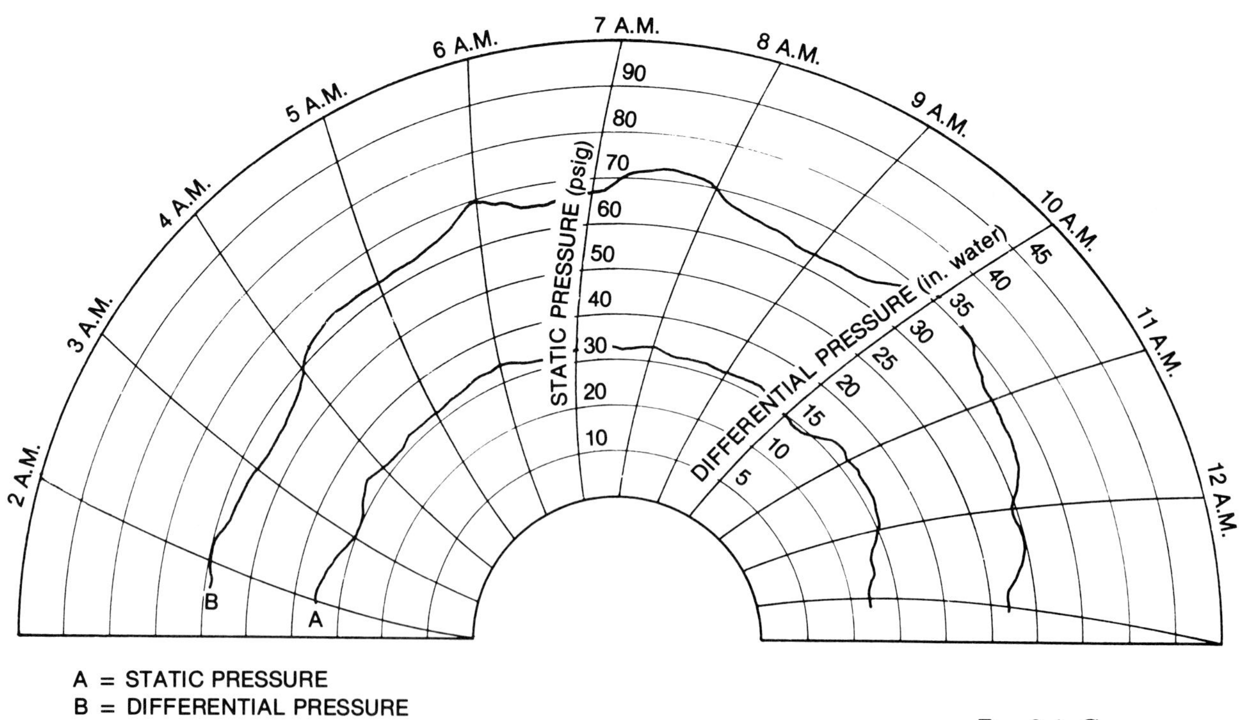

Fig. 8.4. Gas measurement chart (linear)

PRACTICE PROBLEMS

1. A well with an initial shut-in bottomhole pressure of 2,000 psi is tested and found to produce 185 bpd at a stabilized flowing bottomhole pressure of 1,750 psi. What is the productivity index (the barrels per day per psi of pressure drawdown) of the well? ____________

2. How much horsepower is required to drive a duplex double-acting pump that is designed to deliver 4,500 bbl of oil per day at a pressure of 150 psi? ____________

3. Suppose the amount of chemical needed in a bottle test is 0.6 cc per 100 cc of emulsion. If you have a 2% solution, how many pints of treating fluid should you use for each 100 barrels of raw crude? ____________

4. Using an orifice flow constant of 1,000, find the rate of flow (in Mcf/hr) at 2:00 A.M. on the chart shown in figure 8.4. Use 14.65 for atmospheric pressure. ____________

5. Find the volume of gas passing a meter between 8 A.M. and 8 P.M. if the differential pressure averages 100 inches of water and the static pressure averages 150 psig. (Graduations on the chart are from 0 to 500 psig.) The orifice flow constant is 1,000. Use 14.65 psi for atmospheric pressure. ____________

Pipeline

PIPE BUOYANCY

In the construction of underwater pipeline crossings and in laying pipe offshore, it is often necessary to coat the line pipe with concrete in order to overcome the buoyancy of the water. The result is that the pipe gains *negative* buoyancy; that is, it becomes heavier than the water it displaces and sinks. The formula for calculating negative buoyancy is

$$b = W + (-0.34d^2 + 2.78dt + 2.78t^2) \qquad \text{(Eq. 32)}$$

where

b = negative buoyancy, or effective weight of pipe in fresh water, in pounds per foot

W = weight of uncoated pipe, in pounds per foot

d = nominal diameter of pipe, in inches

t = thickness of concrete coating, in inches.

Table 8.4
Negative Pipe Buoyancy

Nominal Pipe Diameter (inches)	Thickness of Concrete Coating (inches, based on 190 lb/cu ft concrete)												
	0″	¼″	½″	¾″	1″	1¼″	1½″	1¾″	2″	2¼″	2½″	2¾″	3″
8⅝″	−25.3	−19.1	−12.6	−5.7	1.5	9.1	17.0	25.2	33.8	42.8	52.1	61.7	71.7
10¾″	−39.4	−31.7	−23.7	−15.3	−6.7	−2.3	11.7	21.2	31.5	42.0	52.7	63.9	75.3
12¾″	−55.2	−46.1	−36.8	−27.0	−17.0	−6.6	4.3	15.3	26.8	38.7	50.8	63.3	76.3
14″	−66.6	−56.7	−46.4	−35.8	−24.9	−12.7	−1.9	10.0	22.3	35.0	48.1	61.4	75.2
16″	−87.0	−75.7	−64.1	−52.1	−39.8	−27.1	−14.0	−0.8	13.0	27.1	41.6	56.2	71.3
18″	−110.0	−97.3	−84.3	−70.9	−57.2	−43.2	−28.7	−14.0	1.1	16.6	32.4	48.5	65.0
20″	−136.0	−121.9	−107.5	−92.8	−77.6	−62.3	−46.4	−30.3	−13.9	1.1	20.3	37.9	55.8
22″	−164.3	−148.8	−133.1	−116.9	−100.4	−83.7	−66.4	−49.0	−31.1	−12.7	5.9	24.7	43.8
24″	−195.8	−178.9	−161.7	−144.2	−126.3	−108.1	−89.5	−70.6	−51.3	−31.7	−11.7	8.6	29.2
26″	−229.5	−211.2	−192.6	−173.7	−154.5	−134.9	−114.8	−94.6	−73.9	−52.8	−31.4	−9.7	12.4
28″	−266.3	−246.7	−226.8	−206.5	−185.9	−165.0	−143.7	−121.9	−100.0	−77.4	−54.9	−32.0	−8.3
30″	−306.0	−285.0	−263.6	−241.9	−219.8	−197.6	−174.7	−151.6	−128.1	−104.1	−80.2	−55.8	−31.0
32″	−348.0	−325.6	−302.9	−279.7	−256.2	−232.6	−208.5	−184.0	−159.0	−133.9	−108.5	−82.6	−56.3
34″	−393.5	−369.7	−348.6	−321.1	−296.3	−271.2	−245.6	−219.9	−193.6	−167.1	−140.2	−113.0	−85.5
36″	−440.0	−414.8	−389.3	−363.4	−337.2	−310.7	−283.7	−256.5	−228.9	−200.9	−172.6	−144.0	−115.0

Values of the expression $(-0.34d^2 + 2.78dt + 2.78t^2)$ are tabulated in table 8.4.

To find the buoyancy of the pipe, given pipe size, weight per foot, and concrete thickness, algebraically add the table value to W. If the resulting value of b is greater than zero, the pipe will sink to the bottom under the indicated conditions. If the resulting value of b is less than zero (negative), the pipe will tend to float, and b becomes a measure of the buoyant force of the water.

Example Problem: Find the negative buoyancy resulting from a 3″ coating of concrete on an 18-inch pipe weighing 82.06 lb/ft.

Solution: In table 8.4 find 3″ in the top boxheads. Follow this column down until you intersect the column whose stub is 18″. The value here is 65.0. Since this is positive, add it to 82.06, the weight of the pipe, and obtain 147.06, the effective weight of the pipe in water, in pounds per foot.

Example Problem: In laying an underwater pipeline crossing, the pipe must be weighted with 190 lb/ft^3 of concrete to achieve a negative buoyancy of 14 lb/ft. The pipe has a nominal diameter of 26″ and a weight of 119.44 lb/ft. Find the thickness of concrete needed to achieve the desired negative buoyancy.

Solution: From equation 32,

$$b = W + \text{reading from table 8.4.}$$

Substituting the known data and transposing the equation,

$$14 = 119.44 + \text{table reading}$$

$$\text{table reading} = 14 - 119.44 = -105.44.$$

Referring to table 8.4, find 26″ in the stub (at left) and follow it across to the number nearest −105.44. This number is −114.8 and is found in the column headed 1½″. Then 1½″ is the required concrete thickness.

PUMPING

In dealing with pumps, it is common to express pressures in terms of the height of a column of water or other fluid (such as mercury) that such a pressure could support, and this height is usually called *hydrostatic head.*

Converting pressure to head and head to pressure:

$$P = 0.433\, H \times SG \qquad \text{(Eq. 33)}$$

$$H = P \div (0.433 \times SG) \qquad \text{(Eq. 34)}$$

where

P = pressure, in psig
H = height of fluid head, in feet
SG = specific gravity of fluid
0.433 = hydrostatic pressure of a vertical column of water, in psi/ft.

Example Problem: What head of water is equivalent to a pressure of 600 psig?

Solution: Use equation 34 and substitute known values:

$P = 600$; specific gravity of water = 1.00.
$H = 600 \div (0.433 \times 1) = 1{,}385.68$ ft.

Example Problem: Express a head of 13 feet of mercury in psig. Specific gravity of mercury is 13.6.

Solution: Use equation 33 and substitute:

$H = 13'$; $SG = 13.6$

Then

$P = 0.433 \times 13 \times 13.6 = 76.6$ psig.

Pump horsepower:

The pump horsepower needed on a pipeline is found from the formula:

$$hp = \frac{R_f \times (P_d - P_s)}{2{,}450 \times E_p} \qquad \text{(Eq. 35)}$$

where

R_f = rate of flow, in bbl/hr
P_d = discharge, or maximum line, pressure, in psig
P_s = suction pressure, in psig
E_p = percentage of pump efficiency, expressed as a factor.

OIL PIPELINE DESIGN

The design of an oil pipeline is based on the consideration of many factors, among which terrain, oil viscosity, oil gravity, pipe size, and desired rate of flow are some of the more important. To illustrate the process by which these factors are evaluated, consider a pipeline to be built under the following conditions:

elevation at origin = 400′
elevation at destination = 3,875′

distance = 100 mi
pipe size = 12"
oil viscosity = 70 SSU
oil gravity = 40°API (sp gr = 0.825)
desired rate of flow = 2,000 bbl/hr
maximum line pressure = 700 psig
suction pressure at each station = 25 psig
pump efficiency = 79%.

The first step is to find the loss in pressure along the line due to friction in the pipe. In figure 8.5 locate the curves dealing with 12" pipe, and the one curve among these that specifies a viscosity of 70 Saybolt Seconds Universal (SSU). On the lower axis, locate the line representing 2,000 bbl/hr; follow this line up to the intersection with the 12"-70 SSU curve. From this intersection, move directly across to the left side of the graph and read "13.5 psi per mile" on the pressure loss axis. This frictional pressure loss is also called the hydraulic gradient. By using equation 33, this loss is converted to a head of 37.8 ft of oil per mile. Then the total pressure loss for the 100-mile line would be

$$37.8 \times 100 = 3{,}780 \text{ ft of head.}$$

Also, the destination of the oil is at elevation 3,875 ft, and the origin is at elevation 400 ft, a difference of 3,475 ft to be overcome. Then the total head to be overcome is

$$3{,}780 + 3{,}475 = 7{,}255 \text{ ft,}$$

which is equivalent to 2,590 psi. One station could supply all of this pressure, but it was specified that the maximum line pressure should not exceed 700 psi. Then four stations would supply the needed pressure with a little to spare.

To determine the locations of these stations, plot a profile of ground elevations along the pipeline route, as shown in figure 8.6. Convert 700 psi (maximum line pressure) to head in feet of oil, 1,960 ft. From the upper end of this line, draw a line sloping to the right at the rate of 37.8 ft per mile, the hydraulic gradient. The place where this line comes to 70 ft above the profile (70 ft is the head of oil equal to the desired suction pressure of 25 psi) is the projected location of station *B*. Repeat the procedure with a 1,960-ft vertical line from station *B* and a line sloping right to station *C*, and so on. It can be seen from the figure that the oil arriving at the tanks at station *E*, the destination, still has several hundred feet of head. Since this head is not necessary just to fill the tanks, the pressure required from the pumps at station *D* can be reduced by this amount.

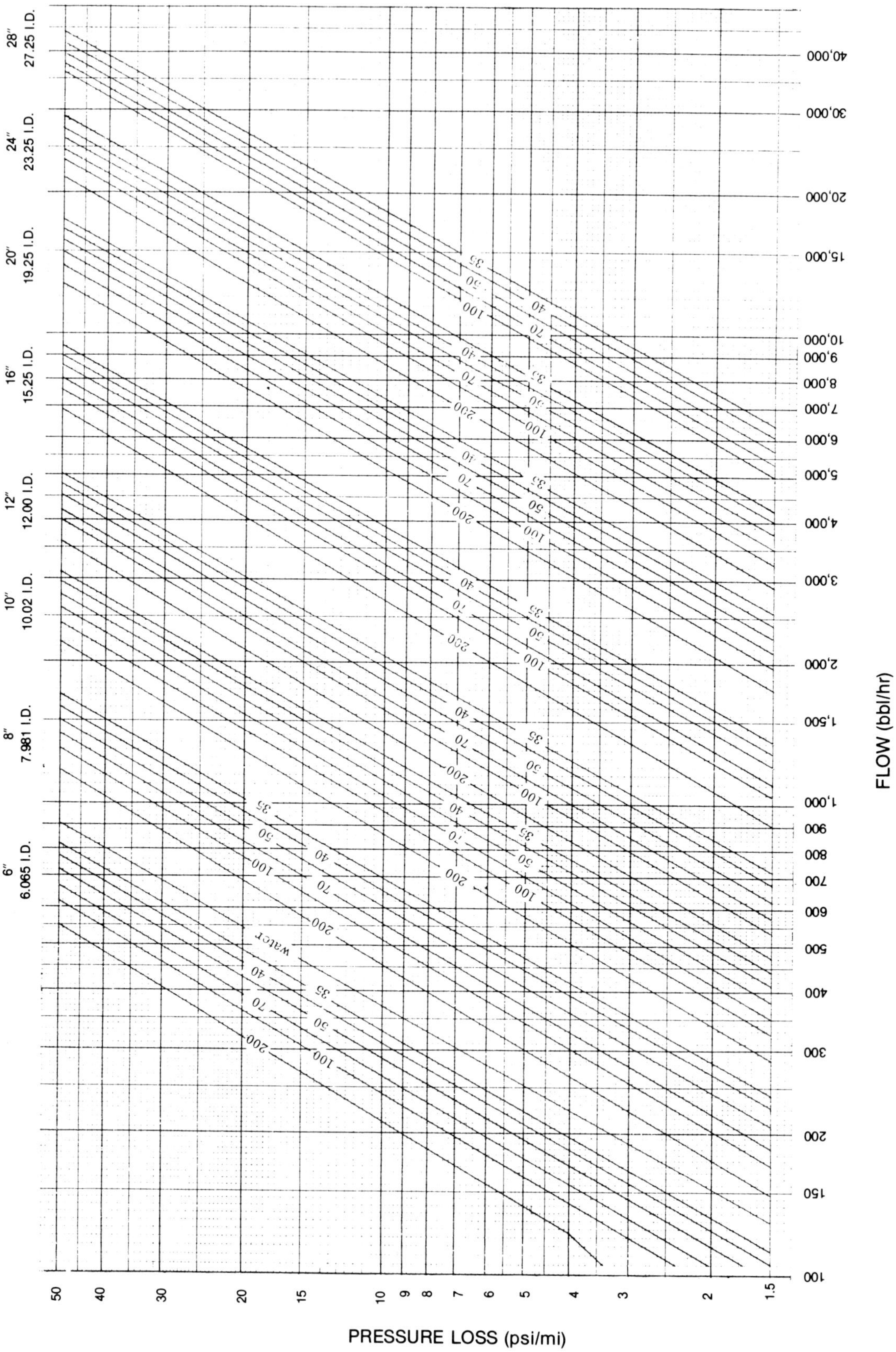

Fig. 8.5. Flow of crude petroleum in pipelines

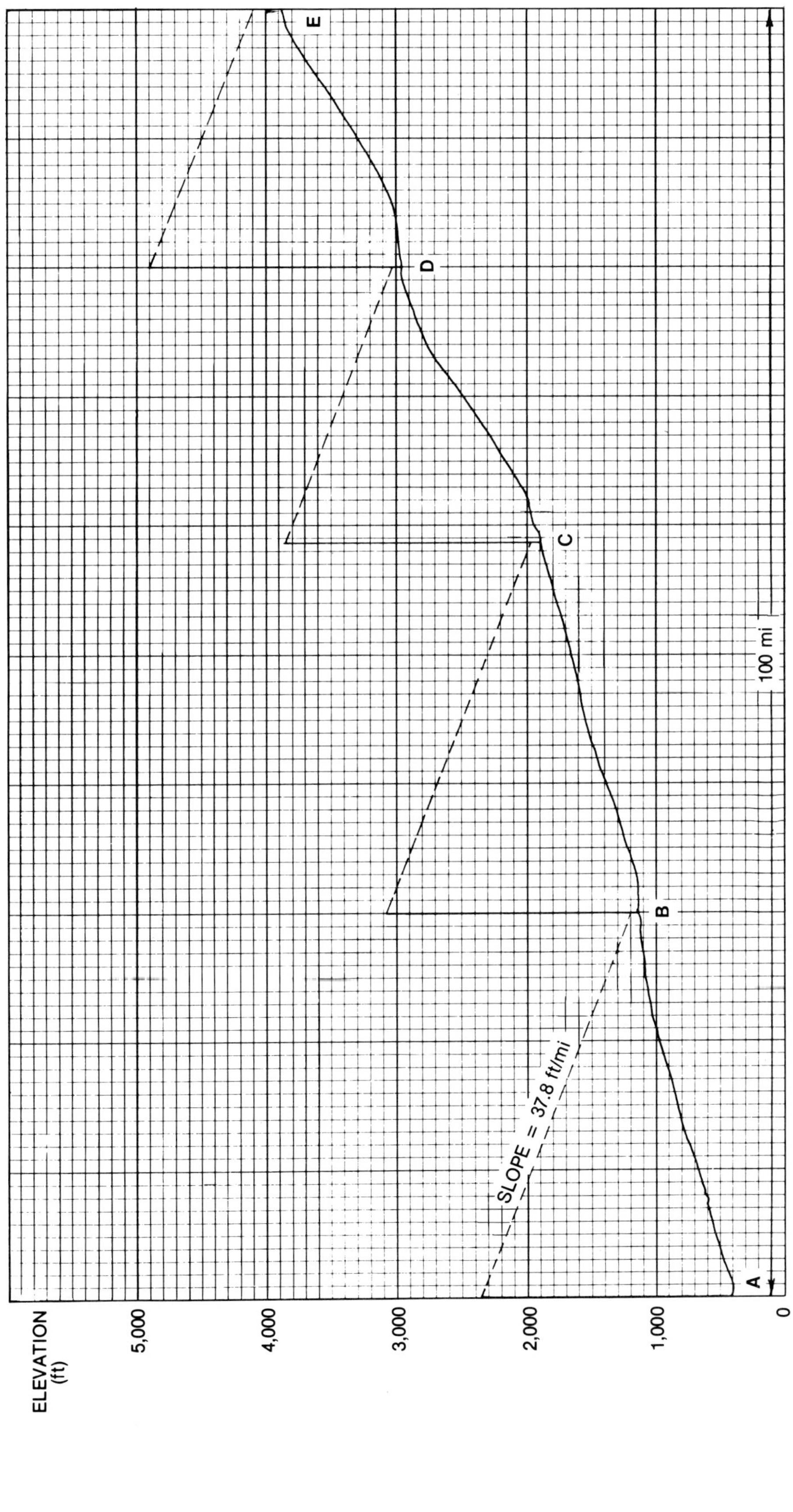

Fig. 8.6. Profile method of calculating pipeline pump station spacing

Using equation 35, the pump horsepower needed on this pipeline would be

$$\frac{2{,}000 \times (700 - 25)}{2{,}450 \times 0.79} = 697.49 \text{ hp.}$$

PRACTICE PROBLEMS

1. What head of water is equivalent to a pressure of 750 psig? ____________

2. Find the pump horsepower needed on a pipeline whose desired rate of flow is 1,500 bbl/hr, discharge pressure is 650 psig, suction pressure is 25 psig, and pump efficiency is 80%. ____________

3. What will be the negative buoyancy of a 20" pipe weighing 95 lb/ft if a 2" coating of concrete is used? ____________

4. A 14" diameter pipe is being used in an underwater pipeline. The uncoated pipe weighs 65 lb/ft. How thick does the cement coating have to be in order to give the pipe an effective weight in water of 18 lb/ft? ____________

5. A 200-mile pipeline loses 35.6 ft of head per mile due to friction. Its elevation at origin is 2,000 ft and at destination, 3,685 ft. What is its total pressure loss in feet of head? ____________

Refining

WEIGHTED AVERAGE

An understanding of weighted averages is necessary in refining problems. For instance, given the data on fuel gas shown in table 8.5, find the net heating value (*LHV*) of the fuel gas.

Table 8.5
Fuel Gas Data

Component	Heating Value of Component (Btu/ft^3)	%Component in Fuel Gas
Methane	910	94
Ethane	1,630	5
Propane	2,350	1

Recall the formula for weighted averages from chapter 3:

$$A = \frac{(N_a \times V_a) + (N_b \times V_b) + (N_c + V_c) \ldots}{a + b + c \ldots} \qquad \text{(Eq. 36)}$$

where

A = weighted average
N = number of items
V = value of each item
a, b, c = items.

The weighted average, or in this case the net heating value of the fuel gas, would be

$$LHV = \frac{(94 \times 910) + (5 \times 1{,}630) + (1 \times 2{,}350)}{94 + 5 + 1} = 960 \text{ Btu/ft}^3.$$

HEAT TRANSFER

To find the amount of heat required to raise the temperature of a volume of liquid, use this formula:

$$Q = WC_p\,(T_2 - T_1) \qquad \text{(Eq. 37)}$$

where

Q = amount of heat, in Btu
W = weight of liquid, in pounds
C_p = specific heat of liquid, in Btu/lb/°F
T_2 = final temperature, in °F
T_1 = initial temperature, in °F.

Example Problem: How much heat is required to heat 100 gal of water from 80°F to 160°F? The specific heat of water is 1.0 Btu/lb/°F.

Solution: The weight of 100 gal of water is

100 gal × 8.33 lb/gal = 833 lb.

Then, using equation 37 and substituting:

Q = 833 × 1.0 Btu/lb/°F × (160 − 80)°F = 66,700 Btu.

If the rate of flow is 100 gpm, then the hourly heat requirement is 66,700 Btu × 60 = 4 MM Btu.

Example Problem: How much heat is required to heat 5,000 bbl of 40°API crude oil from 80°F to 250°F? The specific heat of this oil is 0.47 Btu/lb/°F at 80° and 0.57 Btu/lb/°F at 250°.

Solution: The specific gravity of 40°API oil is 0.825. The weight of 5,000 bbl would be:

5,000 bbl × 0.825 × 8.33 lb/gal × 42 gal/bbl = 1,445,000 lb.

The average C_p is (0.47 + 0.57) ÷ 2 = 0.52 and

$$Q = 1{,}445{,}000 \text{ lb} \times 0.52 \text{ Btu/lb/°F} \times (250 - 80)\text{°F} = 127.7 \text{ MM Btu.}$$

If the rate of flow is 5,000 bpd, then the hourly heat requirement will be 127.7 MM Btu ÷ 24 = 5,350,000 Btu.

FUEL CONSUMPTION

To find fuel costs, use the formula:

$$R = Q \div (E \times LHV) \qquad \text{(Eq. 38)}$$

where

R = fuel consumption, in ft^3
Q = heat requirement, in Btu
E = heater efficiency, expressed as a decimal
LHV = net heating value of gas, in Btu/ft^3.

Then

fuel cost = consumption × value.

Example Problem: If a gas having an *LHV* of 960 Btu/ft^3 costs $2.06 per Mcf, what would be the daily cost of heating the oil in the previous problem (flowing at the rate of 5,000 bpd and requiring 5,350,000 Btu) with a 60% efficient heater?

Solution: Substituting in equation 38:

$$R = 127.7 \text{ MM Btu} \div (0.60 \times 960 \text{ Btu/ft}^3) = 221{,}700 \text{ ft}^3.$$

Then the fuel cost would be:

221,700 × $2,06/1,000 = $456.70.

MATERIAL BALANCE

In working with plant and refinery processes, it is often necessary to calculate the amount of product from a particular process, using known data on feeds and other products. The following example illustrates how this calculation is done.

Example Problem: Using the process and data given in figure 8.7, determine the daily production of bottoms.

Solution: From the data on input and output, convert all to pounds per hour.

Feed: 3,000 bbl/day × (1 day ÷ 24 hr) × 42 gal/bbl × 8.33 lb/gal × 0.946 = 39,700 lb/hr (*input*)

Steam: 2,000 lb/hr (*input*)

Overhead: 280 ft^3/min × 60 min/hr × 0.1835 lb/ft^3 = 3,080 lb/hr (*output*)

Water: 2,000 lb/hr (*output*)

Sidestream: 40 gal/min × 60 min/hr × 8.33 lb/gal × 0.825 = 16,500 lb/hr (*output*)

Bottoms?

Then a balance sheet shows:

Feed	39,700	
Steam	2,000	
Total Input	41,700	lb/hr
Overhead	3,080	
Water	2,000	
Sidestream	16,500	
Total Output	21,580	lb/hr + bottoms

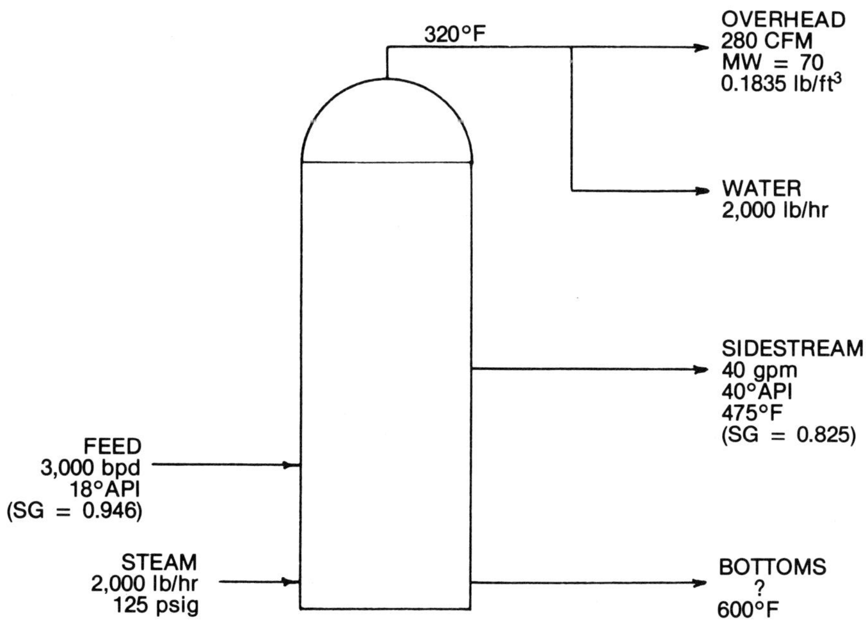

Fig. 8.7

Since the input and output should be equal,

bottoms + 21,580 = 41,700

bottoms = 41,700 − 21,580 = 20,120 lb/hr.

And daily production would be

20,120 lb/hr × 24 hr × (1 ton ÷ 2,000 lb) = 241 tons/day.

PRACTICE PROBLEMS

1. A barrel of liquid (42 gal) having a specific gravity of 0.9 required 32,000 Btu to raise the temperature from 75°F to 200°F. What is the average specific heat of the liquid? ____________

2. An oilfield heater has throughput of 150 bbl/hr and consumes 3,500 ft^3 of gas per hour to increase the temperature from 65°F to 175°F. The gas has a heating value of 1,000 Btu/ft^3. The throughput product has a specific heat of 0.6 and specific gravity of 0.82. What is the efficiency of the heater in percent? ____________

SELF-TEST
8. Advanced Oil Industry Applications

Multiply each problem answered correctly by ten to arrive at your percentage of competency.

Round off your answers to two places. Use information given in the appendices as well as chapter 8 to solve these problems.

1. Solve for mud weight if the hydrostatic pressure is 2,330 psi and the true vertical depth of the well is 4,000 ft.

   ____

2. Air pressure is admitted into a closed container holding mercury as shown in the figure below. The distance between the mercury levels is 25 inches. What is the pressure in the container in psig? Mercury has a specific gravity of 13.6 (846 lb/ft^3), and air has a specific gravity of 0.0012 (0.075 lb/ft^3).

   ____

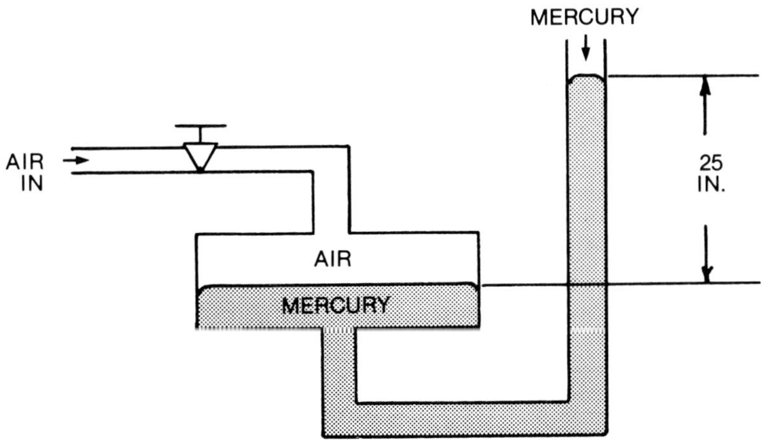

3. What is the hydrostatic pressure of a well whose true vertical depth is 3,452 ft with a mud weight of 9 ppg?

   ____

4. Using an orifice flow constant of 1,000, find the MMcf/day from a well whose differential pressure averages 100 inches of water and whose average static pressure is 150 psig. Use 14.65 rather than 14.7 as the psia factor.

   ____

5. The volume of a gasoline tank is 3,500 cubic feet. What is the weight of the gasoline it will contain when full?

____

6. In batching products through a pipeline, the rate of travel of the fluid through a 130-mile pipeline is 2.5 mph. If a new batch is started at 10:00 A.M. on Monday, when will it first arrive at the terminal end?

____

7. A pump is forcing water into an elevated reservoir tank. The pump must exert enough pressure to offset that of the column of water in the tank and in the pipe leading to it. Assume total vertical height from pump to surface of water in the tank is 150 feet. How much pressure must the pump produce to overcome the effects of hydrostatic pressure due to the water level height?

____

8. A tank has 2,000 cubic feet of gas at 75 psig and 78°F. If the pressure is changed to 80 psig and the temperature drops to 72°F, what will be the volume of the gas?

____

9. A certain petroleum product has a specific head capacity of 0.6 Btu/lb/°F and a specific gravity of 0.85. The product is flowing through a heater at 100 bbl/hr. Its temperature must be increased from 75°F to 180°F. How much heat energy must the heater supply in Btu/hr?

____

10. In problem 9, how much gas will be required for the heater if the gas has a heating value of 1,010 Btu/ft^3, and the heater has an efficiency of 65%? Give your answer in cubic feet.

____

APPENDIX A
Commonly Used Formulas

A-1. PROPORTION AND RATIO

Direct proportion:

$$a:b = c:d, \text{ or } ad = bc$$

Inverse proportion:

$$a:d = b:c, \text{ or } ac = bd$$

where

a = small, or first term
b = small, or second term
c = large, or third term
d = large, or fourth term.

Pulley ratio:

$$S_A/S_B = d_B/d_A$$

where

S_A = speed of pulley A
S_B = speed of pulley B
d_A = diameter of pulley A
d_B = diameter of pulley B.

Gear ratio:

$$S_A \times T_A = S_B \times T_B$$

where

S_A = speed of gear A
T_A = number of teeth in gear A
S_B = speed of gear B
T_B = number of teeth in gear B.

A-2. LENGTH, AREA, AND VOLUME

Board feet:

$$bd\,ft = \frac{t \times w \times l}{12}$$

where

$bd\,ft$ = board feet in lumber
t = thickness of board, in inches
w = width of board, in inches
l = length of board, in feet.

Circle:

$$C = \pi d$$
$$C = \pi 2r$$
$$d = C/\pi$$
$$d = 2r$$
$$A = \pi r^2$$
$$r = \sqrt{A/\pi}$$

where

C = circumference
d = diameter
r = radius
π = 3.1416
A = area.

Cone, right circular:

$$L = \tfrac{1}{2}sC$$
$$V = \tfrac{1}{3}(Bh)$$
$$V = \tfrac{1}{3}(\pi r^2 h)$$

where

L = lateral surface
s = slant height
V = volume
B = area of base
r = radius of base
h = height, or altitude
π = 3.1416
C = circumference of base.

Cube:

$$V = l^3$$

where

V = volume
l = length of one side.

Cylinder, right:

$$L = 2\pi rh$$
$$L = hC$$
$$T = 2\pi r(r + h)$$
$$V = Bh$$
$$V = \pi r^2 h$$

where

L = lateral surface
T = total surface area
r = radius of base
h = altitude
C = circumference of base
π = 3.1416
V = volume
B = area of base.

Ellipse:

$$A = \pi ab$$

where

π = 3.1416
a = one-half the major axis
b = one-half the minor axis.

Elliptical solid:

$$V = \pi abh$$

where

a = one-half the major axis
b = one-half the minor axis
h = height
π = 3.1416.

Frustum of cone or pyramid:

$$V = \tfrac{1}{3}h(B + B' + \sqrt{BB'})$$

where

V = volume
h = height of frustum
B = area of cone or pyramid base
B' = area of section base.

Parallelogram:

$$A = bh, \text{ or } lw$$

where

b = base
h = height
l = length
w = width.

Polygon, regular:

$$A = \tfrac{1}{2}aP$$

where

A = area
a = altitude, or apothem
P = perimeter.

Pyramid, regular:

$$L = \tfrac{1}{2}sP$$
$$V = \tfrac{1}{3}(Bh)$$

where

L = lateral area
s = slant height
P = perimeter
V = volume
B = area of base
h = height, or altitude.

Rectangle:

$$A = bh, \text{ or } lw$$
$$P = 2l + 2w$$

where

A = area
b = base
h = height
l = length
w = width
P = perimeter.

Rectangular solid:

$$V = lwh$$

where

V = volume
l = length
w = width
h = height.

Sphere:

$$V = \tfrac{4}{3}\pi r^3$$

where

V = volume
r = radius
π = 3.1416.

Square:

$$P = 4 \times l$$
$$A = l^2, \text{ or } bh$$

where

l = length of one side
b = base
h = height.

Trapezoid:

$$A = \tfrac{1}{2}h(b + b')$$

where

A = area
h = height
b = bottom base
b' = top base.

Triangle:

$$A = \tfrac{1}{2}bh$$

where

A = area
b = base
h = height.

A-3. TRIGONOMETRIC FUNCTIONS

In the following functions involving a right triangle,

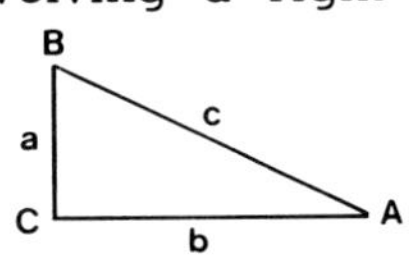

A = angle opposite side a
B = angle opposite side b
C = right angle
c = hypotenuse, or side opposite right angle C
a = side adjacent to angle B, opposite angle A
b = side adjacent to angle A, opposite angle B.

Sine:

sin A = a/c
sin B = b/c.

Cosine:

cos A = b/c
cos B = a/c.

Tangent:

tan A = a/b
tan B = b/a.

Cosecant:

csc A = c/b
csc B = c/a.

Secant:

sec A = c/a
sec B = c/a.

Cotangent:

cot A = b/a
cot B = a/b.

A-4. ELECTRICITY

Ohm's law:

$$I = E/R$$
$$E = IR$$
$$R = E/I$$

where

I = current, in amperes
E = voltage, in volts
R = resistance, in ohms.

Series circuit current flow:

$$I = \frac{E}{R_1 + R_2 + R_3 + \ldots}$$

where

I = current, in amperes
E = voltage, in volts
$R_1, R_2, R_3, \ldots$ = resistors, in ohms

Parallel circuit total current:

$$I = E/R_1 + E/R_2 + E/R_3 + \ldots$$

Parallel circuit resistance:

$$R = 1/R_1 + 1/R_2 + 1/R_3 + \ldots$$

Power:

$$P = E \times I$$
$$P = I^2R$$

where

P = power, in watts
E = voltage, in volts
I = current, in amperes.

A-5. DENSITY AND SPECIFIC GRAVITY

Specific gravity, or relative density:

$$SG = D_s \div D_w$$

$$D_s = D_w \times SG$$

where

SG = specific gravity

D_s = density of substance, or weight per volume of substance, in lb/ft^3, kg/m^3, or g/cm^3.

D_w = density of pure water, or weight per volume of water, in lb/ft^3, kg/m^3, or g/cm^3.

Baumé scale for heavy liquids:

$$°Baumé = 145 - 145/SG.$$

Baumé scale of light liquids:

$$°Baumé = 140/SG - 130.$$

°API gravity:

$$°API = 141.5/SG - 131.5.$$

A-6. POWER

Pound-force:

$$lb_f = lb_m \times 32.15$$

where

lb_f = pound-force

lb_m = pound-mass

32.15 = a constant, in feet per second squared.

Horsepower:

$$hp = ft\text{-}lb/min \div 33{,}000$$

where

hp = horsepower

ft-lb = foot-pounds of work

min = time, in minutes

33,000 = a constant.

A-7. PRESSURE-VOLUME-TEMPERATURE

Converting gauge pressure to absolute pressure:

$$psia = psig + 14.65$$

$$psig = psia - 14.65$$

where

psia = absolute pressure, in pounds per square inch

psig = gauge pressure, in pounds per square inch

14.65 = a constant.

Charles's law:

$$P_2/P_1 = T_2/T_1$$

where

P_2 = absolute pressure after a change in temperature

P_1 = initial absolute pressure

T_2 = absolute temperature after a change in pressure

T_1 = initial absolute temperature.

Note: Volume of gas must remain the same.

Boyle's law:

$$P_1V_1 = P_2V_2$$

where

P_1 = initial absolute pressure

V_1 = initial volume of gas

P_2 = absolute pressure after a change in volume

V_2 = volume of gas after a change in pressure.

Note: Temperature must remain the same.

Ideal gas law:

$$P_1V_1 \div P_2V_2 = T_1 \div T_2$$

where

P_1 = original absolute pressure

P_2 = final absolute pressure

V_1 = original volume

V_2 = final volume

T_1 = original absolute temperature

T_2 = final absolute temperature.

APPENDIX B
Reference Tables

B-1. DECIMAL EQUIVALENTS

Fraction	Decimal Equivalent	Fraction	Decimal Equivalent
1/64	0.015625	33/64	0.515625
1/32	0.03125	17/32	0.53125
3/64	0.046875	35/64	0.546875
1/16	0.0625	9/16	0.5625
5/64	0.078125	37/64	0.578125
3/32	0.09375	19/32	0.59375
7/64	0.109375	39/64	0.609375
1/8	**0.125**	**5/8**	**0.625**
9/64	0.140625	41/64	0.640625
5/32	0.15625	21/32	0.65625
11/64	0.171875	43/64	0.671875
3/17	0.1875	11/16	0.6875
13/64	0.203125	45/64	0.703125
7/32	0.21875	23/32	0.71875
15/64	0.234375	47/64	0.734375
1/4	**0.25**	**3/4**	**0.75**
17/64	0.265625	49/64	0.765625
9/32	0.28125	25/32	0.78125
19/64	0.296875	51/64	0.796875
5/16	0.3125	13/16	0.8125
21/64	0.328125	53/64	0.828125
11/32	0.34375	27/32	0.84375
23/64	0.359375	55/64	0.859375
3/8	**0.375**	**7/8**	**0.875**
25/64	0.390625	57/64	0.890625
13/32	0.40625	29/32	0.90625
27/64	0.421875	59/64	0.921875
7/16	0.4375	15/16	0.9375
29/64	0.453125	61/64	0.953125
15/32	0.46875	31/32	0.96875
31/64	0.484375	63/64	0.984375
1/2	**0.5**	**1**	**1**

B-2. SQUARES, CUBES, SQUARE ROOTS, AND CUBE ROOTS OF NUMBERS

No.	Square	Cube	Sq. Root	Cube Root	No.	Square	Cube	Sq. Root	Cube Root
1	1	1	1.00	1.00	26	676	17,576	5.10	2.96
2	4	8	1.41	1.26	27	729	19,683	5.20	3.00
3	9	27	1.73	1.44	28	784	21,952	5.29	3.04
4	16	64	2.00	1.59	29	841	24,389	5.38	3.07
5	25	125	2.24	1.71	30	900	27,000	5.48	3.11
6	36	216	2.45	1.82	31	961	29,791	5.57	3.14
7	49	343	2.65	1.91	32	1,024	32,768	5.66	3.17
8	64	512	2.83	2.00	33	1,089	35,937	5.74	3.21
9	81	729	3.00	2.08	34	1,156	39,304	5.83	3.24
10	100	1,000	3.16	2.15	35	1,225	42,875	5.92	3.27
11	121	1,331	3.32	2.22	36	1,296	46,656	6.00	3.30
12	144	1,728	3.46	2.29	37	1,369	50,653	6.08	3.33
13	169	2,197	3.61	2.35	38	1,444	54,872	6.16	3.36
14	196	2,744	3.74	2.41	39	1,521	59,319	6.24	3.39
15	225	3,375	3.87	2.47	40	1,600	64,000	6.32	3.42
16	256	4,096	4.00	2.52	41	1,681	68,921	6.40	3.45
17	289	4,913	4.12	2.57	42	1,764	74,088	6.48	3.48
18	324	5,832	4.24	2.62	43	1,849	79,507	6.56	3.50
19	361	6,859	4.36	2.67	44	1,936	85,184	6.63	3.53
20	400	8,000	4.47	2.71	45	2,025	91,125	6.71	3.56
21	441	9,261	4.58	2.76	46	2,116	97,336	6.78	3.58
22	484	10,648	4.69	2.80	47	2,209	103,823	6.86	3.61
23	529	12,167	4.80	2.84	48	2.304	110,592	6.93	3.63
24	576	13,824	4.90	2.88	49	2,401	117,649	7.00	3.66
25	625	15,625	5.00	2.92	50	2,500	125,000	7.07	3.68

B-3. METRIC PREFIXES

Numerical Value	Name	Symbol	Numerical Value	Name	Symbol
*10^1	deca	da	*10^{-1}	deci	d
*10^2	hecto	h	*10^{-2}	centi	c
10^3	kilo	k	10^{-3}	milli	m
10^6	mega	M	10^{-6}	micro	μ
10^9	giga	G	10^{-9}	nano	n
10^{12}	tera	T	10^{-12}	pico	p
10^{15}	peta	P	10^{-15}	femto	f
10^{18}	exa	E	10^{-18}	atto	a

*Not encouraged for use in SI.

B-4. CONVENTIONAL AND METRIC (SI) UNITS OF MEASUREMENT

Quantity	Conventional Unit	Conventional Symbol	Metric (SI) Unit	Metric (SI) Symbol
Fundamental Quantities				
Length	foot	ft	metre	m
Mass	pound	lb	kilogram	kg
Time	second	sec	second	s
Temperature	degree Fahrenheit degree Rankine	°F °R	degree Celsius kelvin	°C K
Electrical current	ampere	A	ampere	A
Amount of substance	mole	mol	mole	mol
Luminous intensity	candela	cd	candela	cd
Derived Quantities				
Acceleration	feet per second per second	ft/sec^2	metre per second per second	m/s^2
Area	square feet	ft^2	square metre	m^2
Density	pound per cubic foot	lb/ft^3	kilogram per cubic metre	kg/m^3
Electrical potential	volt	V	volt	V
Electrical resistance	ohm	Ω	ohm	Ω
Electrical capacitance	farad	F	farad	F
Energy	kilowatt-hour	kWh	kilowatt-hour	kWh
Force	pound-force	lb_f	newton	N
Frequency	hertz	Hz	hertz	Hz
Power	foot-pound per second horsepower watt	ft-lb/sec hp W	joule per second watt watt	J/s W W
Pressure	pound-force per square inch	psi	pascal	Pa
Velocity	foot per second	ft/sec	metre per second	m/s
Volume	cubic foot gallon barrel	ft^3 gal bbl	cubic metre	m^3
Work	foot-pound	ft-lb	joule newton-metre	J N·m

B-5. NATURAL TRIGONOMETRIC FUNCTIONS

Angle	Sine	Cosine	Tangent	Angle	Sine	Cosine	Tangent
1°	0.0175	0.9998	0.0175	46°	0.7193	0.6947	1.0355
2°	0.0349	0.9994	0.0349	47°	0.7314	0.6820	1.0724
3°	0.0523	0.9986	0.0524	48°	0.7431	0.6691	1.1106
4°	0.0698	0.9976	0.0699	49°	0.7547	0.6561	1.1504
5°	0.0872	0.9962	0.0875	50°	0.7660	0.6428	1.1918
6°	0.1045	0.9945	0.1051	51°	0.7771	0.6293	1.2349
7°	0.1219	0.9925	0.1228	52°	0.7880	0.6157	1.2799
8°	0.1392	0.9903	0.1405	53°	0.7986	0.6018	1.3270
9°	0.1564	0.9877	0.1584	54°	0.8090	0.5878	1.3764
10°	0.1736	0.9848	0.1763	55°	0.8192	0.5736	1.4281
11°	0.1908	0.9816	0.1944	56°	0.8290	0.5592	1.4826
12°	0.2079	0.9781	0.2126	57°	0.8387	0.5446	1.5399
13°	0.2250	0.9744	0.2309	58°	0.8480	0.5299	1.6003
14°	0.2419	0.9703	0.2493	59°	0.8572	0.5150	1.6643
15°	0.2588	0.9659	0.2479	60°	0.8660	0.5000	1.7321
16°	0.2756	0.9613	0.2867	61°	0.8746	0.4848	1.8040
17°	0.2924	0.9563	0.3057	62°	0.8829	0.4695	1.8807
18°	0.3090	0.9511	0.3249	63°	0.8910	0.4540	1.9626
19°	0.3256	0.9455	0.3443	64°	0.8988	0.4384	2.0503
20°	0.3420	0.9397	0.3640	65°	0.9063	0.4226	2.1445
21°	0.3584	0.9336	0.3839	66°	0.9135	0.4067	2.2460
22°	0.3746	0.9272	0.4040	67°	0.9205	0.3907	2.3559
23°	0.3907	0.9205	0.4245	68°	0.9272	0.3746	2.4751
24°	0.4067	0.9135	0.4452	69°	0.9336	0.3584	2.6051
25°	0.4226	0.9063	0.4663	70°	0.9397	0.3420	2.7475
26°	0.4384	0.8988	0.4877	71°	0.9455	0.3256	2.9042
27°	0.4540	0.8910	0.5095	72°	0.9511	0.3090	3.0777
28°	0.4695	0.8829	0.5317	73°	0.9563	0.2924	3.2709
29°	0.4848	0.8746	0.5543	74°	0.9613	0.2756	3.4874
30°	0.5000	0.8660	0.5774	75°	0.9659	0.2588	3.7321
31°	0.5150	0.8572	0.6009	76°	0.9703	0.2419	4.0108
32°	0.5299	0.8480	0.6249	77°	0.9744	0.2250	4.3315
33°	0.5446	0.8387	0.6494	78°	0.9781	0.2079	4.7046
34°	0.5592	0.8290	0.6745	79°	0.9816	0.1908	5.1446
35°	0.5736	0.8192	0.7002	80°	0.9848	0.1736	5.6713
36°	0.5878	0.8090	0.7265	81°	0.9877	0.1564	6.3138
37°	0.6018	0.7986	0.7536	82°	0.9903	0.1392	7.1154
38°	0.6157	0.7880	0.7813	83°	0.9925	0.1219	8.1443
39°	0.6293	0.7771	0.8098	84°	0.9945	0.1045	9.5144
40°	0.6428	0.7660	0.8391	85°	0.9962	0.0872	11.4301
41°	0.6561	0.7547	0.8693	86°	0.9976	0.0698	14.3007
42°	0.6691	0.7431	0.9004	87°	0.9986	0.0523	19.0811
43°	0.6820	0.7314	0.9325	88°	0.9994	0.0349	28.6363
44°	0.6947	0.7193	0.9657	89°	0.9998	0.0175	57.2900
45°	0.7071	0.7071	1.0000	90°	1.0000	0.0000	∞

B-6. DENSITY/SPECIFIC GRAVITY TABLES

Densities of Some Solids, Liquids, and Gases

Material	lb/ft^3	kg/m^3	g/cm^3 (sp gr)
Solids			
Gold	1,206.2	19,300	19.3
Mercury	846.0	13,500	13.5
Lead	712.5	11,400	11.4
Iron	485.0*	7,700*	7.7*
Aluminum	165.6	2,600	2.6
Wood	50.0*	800*	0.8*
Ice	56.9	900	0.9
Liquids			
Sulfuric acid	125.0	2,000	2.0
Seawater	64.3	1,030	1.03
Fresh (pure) water	62.5	1,000	1.00
Kerosine	50.0	800	0.80
Gasoline	46.8	750	0.75
Gases			
Air	0.075	1.20	0.0012
Oxygen	0.084	1.34	0.00134
Nitrogen	0.0737	1.18	0.0018
Carbon monoxide	0.0734	1.17	0.0017
Hydrogen	0.0053	0.085	0.000085

*Wood and iron have variable densities. The values shown are approximate figures.

Specific Gravity Based on Air

Gas	Sp gr
Air	1.00
Oxygen	1.120
Nitrogen	0.0983
Carbon monoxide	0.979
Hydrogen	0.071
Butane	2.004

APPENDIX C
Conversion Factors

C-1. CONVERSION FACTORS FOR CONVENTIONAL AND METRIC (SI) MEASUREMENTS

To use this table, multiply the number of known units of measure (find in left-hand column) by the conversion factor for the desired unit of measure; for example, to convert 50 barrels to gallons, multiply: 50 × 42 = 2,100 gallons.

Unit of Measure	Conventional Units		Metric Units	
To convert:	*to:*	*multiply by:*	*to:*	*multiply by:*
acre	square foot	43,560	square metre	4,046.875
	square yard	4,840	square kilometre	0.004046875
	square rod	160		
	square mile	0.0015625		
	section	0.0015625		
acre-foot	cubic foot	43,560	cubic metre	1,233.48766
bar	inch of water @ 60°F	407.229	millimetre of water @ 60°F	10,343.6
	inch of mercury @ 32°F	29.5282	millimetre of mercury @ 0°C	750.0187
barrel, U.S. petroleum (bbl)	cubic foot	5.6146	cubic metre	0.1589873
	gallon	42	litre	158.9873
British thermal unit @ 60°F (Btu)	foot-pound	777.97265	joule	1,054.68
	kilowatt-hour	0.00029283	kilowatt-hour	0.00029283
	calorie	251.98		
	horsepower-hour	0.0003927		
calorie (cal)	British thermal unit	0.003968	joule	4.1868
centimetre (cm)	mil	393.70	millimetre	10
	inch	0.3937	metre	0.01
	foot	0.032808		
centipoise (cp)	poise	100	pascal-second	0.001
cubic centimetre (cm^3)	cubic inch	0.061023	cubic millimetre	1,000
	cubic foot	0.000035314	millilitre	1
	gallon	0.00026417		
cubic foot	cubic inch	1,729.98829	cubic centimetre	28,317
	cubic yard	0.037036	cubic metre	0.028317
	gallon	7.48050		
	barrel	0.17811		

Unit of Measure *To convert:*	Conventional Units *to:*	*multiply by:*	Metric Units *to:*	*multiply by:*
cubic foot of water (ft³ of H_2O)	pound of water	62.3		
cubic foot/second (ft^3/sec)	gallon/minute	488.883		
cubic inch ($in.^3$)	cubic foot	0.0005787	cubic centimetre	16.38716
	cubic yard	0.000021434	litre	0.016387
	gallon	0.0043290	millilitre	16.38716
cubic metre (m^3)	cubic foot	35.314445	kilolitre	1
	cubic inch	61,022.9388	litre	1,000
	cubic yard	1.307943	cubic centimetre	1,000,000
	gallon	264.17762		
	barrel	6.28994		
cubic yard (yd^3)	cubic foot	4.80897	cubic metre	0.76456
	gallon	201.97350	kilolitre	0.76456
	barrel	4.80897		
day	minute	0.0006944		
	hour	0.0041666		
	week	7		
	year	365		
degree, angle	minute	60	radian	0.01745
	second	3,600		
degree, temperature (See Appendix C-2.)				
foot/second (ft/sec)	miles/hour	0.68182	metre/second	0.3048
foot-pound force ($ft\text{-}lb_f$)	inch-pound force	12	joule	1.35582
	Btu	0.001286		
gallon of water	pound of water	8.328		
gallon, U.S. liquid (gal)	cubic inch	231	cubic metre	0.0037854
	cubic foot	0.133681		
	barrel	0.0238095		
gram (g)	oz, avoir.	0.0352739	kilogram	1,000
hectare	acre	2.471	square metre	10,000
horsepower (hp)	foot-pound/minute	33,000	watt	745.6999
	foot-pound/second	550		

Unit of Measure	Conventional Units		Metric Units	
To convert:	*to:*	*multiply by:*	*to:*	*multiply by:*
horsepower-hour	Btu	2,546.473	megajoule	2.68452
hour (hr)	second	0.0002777	second	0.0002777
	minute	0.016666		
	day	24		
	year	8,766		
inch (in.)	foot	0.083333	metre	0.0254
	mil	1,000	centimetre	254
			millimetre	25.4
inch of mercury . . . . @ 32°F	inch of water @ 60°F	13.595326	pascal	3,386.389
	pound/square inch	0.491161	millimetre of mer- cury @ 0°C	
	bar	0.0338659		
joule (J)	foot-pound	0.73756	kilowatt-hour	0.0000002778
	watt-second	1	watt-second	1
kilogram (kg)	pound mass, avoir.	2.204622	gram	1,000
			tonne	0.001
kilometre (km).	mile, U.S. statute	0.62137	metre	1,000
kilometre/hour . . . . (km/h)	mile/hour	0.62137	metre/second	0.2777778
kilowatt (kW).	foot-pound/second	737.56	joule/second	3,600,000
	horsepower	1.341		
kilowatt-hour (kWh)	horsepower-hour	1.341	joule	3,600,000
litre	gallon	0.264178	cubic metre	0.001
	cubic inch	61.025		
megajoule (MJ)	horsepower-hour	0.3725	joule	1,000,000
metre (m)	inch	39.370	centimetre	100
	foot	3.280833	kilometre	0.001
mil	inch	0.001	millimetre	0.0254
mile, U.S. statute (mi)	rod	320	kilometre	1.60935
	foot	5,280	metre	1,609.35
	yard	1,760		
miles per hour (mph)	foot/second	1.467	metre/second	0.44704

Unit of Measure *To convert:*	Conventional Units *to:*	*multiply by:*	Metric Units *to:*	*multiply by:*
millibar	bar	1000	pascal	101.356
millimetre of mercury @ 0°C (mm of Hg)	inch of mercury @ 32°F	0.03937	pascal	133.3224
minute (min)	second	0.016666	second	0.016666
	hour	60		
	day	1,440		
newton (N)	pound-force	0.2248		
newton-metre (N·m)	pound-force-foot	0.7375621	joule	1
ohm (Ω)	megohm	0.000001	megohm	0.000001
ounce, U.S. fluid (oz)	cubic inch	1.80469	cubic metre	0.00002957
ounce-force (oz_f)	pound-force	16	newton	0.278
ounce-mass (oz_m)	pound-mass	16	kilogram	0.4535
pascal (Pa)	pound-force/ square inch	0.000145038	newton/square metre	1
pascal-second (Pa·s)	centipoise	0.001	gram/centimetre-second	0.10
	poise	0.10		
poise	centipoise	100	pascal-second	0.10
pound-force (lb_f)			newton	4.448
pound-force/foot			newton/metre	14.59390
pound-force/square inch (psi)	foot of water @ 60°F	2.306787	pascal	6,894.757
	inch of mercury @ 32°F	2.0360	kilopascal	6.894757
	bar	0.068947		
pound-mass (lb_m)	ounce-mass	16	kilogram	0.4536
pound of water	gallon of water	0.120		
	cubic foot	0.016		
radian, angle (rad)	degree	57.29578		
	degree, minute, second	57°17′44.8″		

Unit of Measure *To convert:*	Conventional Units *to:*	 *multiply by:*	Metric Units *to:*	 *multiply by:*
rod (rd)	foot	16.5	metre	5.0292188
	yard	5.5	centimetre	502.921875
	mile, U.S. statute	0.003125		
section	square mile	1	square metre	2,589,988
	acre	640	hectare	258.9988
square foot (ft^2)	square inch	144	square metre	0.0929034
	square yard	0.11111	square centimetre	929.0341
square inch ($in.^2$)	square foot	0.00694444	square centimetre	6.451626
	square mil	1,000,000	square millimetre	645.16258
	circular mil	1,273,240		
square kilometre (km^2)	square mile	0.383101	hectare	100
	acres	247.1044	square metre	1,000,000
square metre (m^2)	square foot	10.7639	square centimetre	10,000
	acres	0.000247104	hectare	0.0001
square mil (mil^2)	circular mil	1.273224	square millimetre	0.000645163
	square inch	0.000001		
square mile, statute (mi^2)	section	1	square metre	2,589,988
	acres	640	square kilometre	2.589988
	square foot	27,878,400	hectare	258.9988
square millimetre (mm^2)	square inch	0.00155	square metre	0.000001
	square mil	1,550.0	square centimetre	0.01
square rod (rd^2)	square foot	272.25	square metre	25.2929
square vara, Texas	square foot	7.716	square metre	0.716843
square yard (yd^2)	square foot	9	square metre	0.836131
ton, U.S. long (tn)	pound-mass	2,240	kilogram	1,016.047
	barrels of oil, 36°API	7.33627	tonne	1.016047
ton, U.S. short (tn)	pound-mass	2,000	tonne	0.90718
tonne, metric SI (t)	pound-mass	2,204.6	kilogram	1,000
	ton, U.S. short	1.01606	gram	1,000,000
	barrel of oil, 36°API	7.454		

Unit of Measure *To convert:*	Conventional Units *to:*	*multiply by:*	Metric Units *to:*	*multiply by:*
township	square mile	36	square kilometre	93.24
vara, Texas	foot	2.777778	metre	0.84667
watt (W)	kilowatt	0.001	kilowatt	0.001
	British thermal unit/hour	3.413772	kilogram-metre/ hour	367.1
	foot-pound/second	0.73756	joule/second	1
	horsepower	0.001341		
watt-hour (Wh)	foot-pound	2,655.22	kilogram-metre	367.1
	kilowatt-hour	0.001	kilowatt-hour	0.001
	horsepower-hour	0.001341	joule	3,600
week (wk)	day	0.1428571		
	year	52		
	hour	0.0059523		
yard (yd)	mile, statute	0.000568182	metre	0.914404
	foot	3	millimetre	914.40360
	inch	36		
	mil	36,000		
year (yr)	month	0.083333		
	day	0.0027397		

C-2. TEMPERATURE CONVERSION FACTORS

To convert:	Conventional *to:*	*factor:*	Metric *to:*	*factor:*
degree Fahrenheit (°F)	°R	°F + 459.67	°C	(°F − 32)/1.8
			K	(°F + 459.67)/1.8
degree Celsius (°C)	°F	°C/1.8 + 32	K	°C + 273.15
	°R	(°R − 491.67)/1.8		
degree Rankine (°R)	°F	°R − 459.67	°C	(°R − 491.67)/1.8
			K	°R/1.8
kelvin (K)	°F	1.8(K − 273.15) + 32	°C	K − 273.15
	°R	1.8K		

ANSWER KEY

Problems involving decimals may have discrepancies in their answers—for example, the problem involving the number of fans needed to change the air every 3 minutes in a given area. Possible answers are 4.053, 4.05, 4, or 5 fans, depending upon the practical application. Four fans will not quite do the job. Also, with 4 fans, there is no standby for emergency. The engineer would probably call for 5 fans. If the specification for a change of air each 3 minutes included a factor for safety, perhaps 4 fans would suffice. Obviously, it is not possible to install 4.053 fans, but that is the solution to the problem to three decimal places. At this point, judgment enters the picture.

Problems involving only whole numbers are precise in the results obtained. Many problems in fractions cannot be worked to a final answer. Rounding off fractions to whole numbers, to one place, two places, and so forth will change the end result. For example, using pi on a calculator instead of 3.1416 will slightly alter an answer. Using rounded-off conversion factors as given in the chapter tables will result in answers different from those solved by using the exact conversion factors given in Appendix C. These answers may vary considerably if large quantities are involved.

The user of this manual should compare his answer with that given, and if there is a difference, he should check his work. If it is not in error, then he might ask, "How can this difference be accounted for?" or "Is my answer satisfactory although perhaps not exact?" If it is satisfactory for the purpose he had in mind from the beginning, then it should be acceptable. However, he should always be able to explain how the difference arose.

Answers to Practice Problems

1. THE NUMBER SYSTEM

Whole Numbers, pp. 10–11:

1. a. 5,000
   b. 110
   c. 30
   d. 10
2. a. 42
   b. 361,000,000
   c. 27,051,289,000
   d. 20,400,502,000,000
3. a. four thousand, six hundred
   b. seventy-eight thousand
   c. eight million, six hundred thousand
   d. eighty billion, seven hundred forty million dollars
4. a. 953
   b. 236
   c. 1,281,150
   d. 6,485,381
5. a. $2,450 per month
   $29,400 per year
   b. 17 hr
   c. 1,080 ft
   d. 4,811 ft

Common Fractions, pp. 16–18:

1. a. 6/64
   b. 12/16
   c. 14/16
   d. 7/8
2. a. 4/8, 1/8, 6/8
   b. 4/32, 2/32, 7/32
   c. 56/64, 19/64, 22/64, 17/64, 4/64, 62/64
   d. 62/64, 60/64, 3/64
3. a. 1/2
   b. 7/8
   c. 1/8
   d. 5/6
4. a. $1\frac{1}{2}$
   b. $3\frac{5}{16}$
   c. 7/32
   d. 43/64
5. a. $5\frac{5}{8}$
   b. 35/160 or 7/32
   c. $1\frac{1}{10}$
   d. 5/18
6. 30 bbl
7. Jones: $28,125
   Smith: $14,062.50
   White: $32,812.50
8. 5/16″
9. 7′$5\frac{9}{16}$″
10. 21/64″

Mixed Numbers, pp. 21–22:

1. 40 hr
2. $17\frac{15}{16}$″
3. $20\frac{29}{32}$″
4. $12\frac{1}{4}$ hr
5. 5/16″
6. $45\frac{3}{8}$″
7. $764\frac{4}{13}$ bbl
8. 10
9. 21
10. 1,100 bbl

Decimal Fractions, pp. 29–30:

1. a. three thousandths
   b. six hundred twenty-five thousandths
   c. one hundred twenty and four hundredths
   d. eight and three thousand, seven hundred, forty-five ten thousandths
2. a. 0.115; 115/1000
   b. 76.76; $76\frac{76}{100}$ or $76\frac{19}{25}$
   c. 5,000.6; $5{,}000\frac{6}{10}$ or $5{,}000\frac{3}{5}$
   d. 3,125.8; $3{,}125\frac{8}{10}$ or $3{,}125\frac{4}{5}$
3. $216.16
4. $570.15
5. 8.928 gal
6. 42.48 hp
7. 53,400 lb
8. $23.03
9. 1.53 in.
10. 1.25 in.

Roots and Powers, p. 33:

1. 28
2. 2,744
3. 36 mi^2
4. 3.2
5. 13,839

2. THE CALCULATOR

p. 54:

1. 117.8
2. 1,391.84
3. 5,035.24
4. 1,548.55
5. 1,872
6. 0.50
7. 548.17
8. 625
9. 408
10. 1,610,069.92
11. 59.45
12. 308 mi
13. $1.29
14. $14.70
15. 1,276.66

3. NUMBER RELATIONS

Percentage, pp. 60–61:

1. $54.60
2. 17
3. $116.40
4. a. Tin—416.50 lb
   b. Copper—28 lb
   c. Antimony—55.50 lb
5. 200
6. $294.40
7. 15%
8. a. Lathe—53.70%
   b. Drill press—8.33%
   c. Shaper—25%
   d. Welding—12.96%
9. 1,282.05 bbl
10. a. 22
    b. $400/wk
    c. 99%
    d. 87½%
    e. 50%

Ratio and Proportion, pp. 65–66:

1. 25 to 1
2. 5 to 3
3. 14 to 1
4. 816 sacks
5. 40 ft

3. NUMBER RELATIONS, CON'T.

6. 50 men
7. 80 mph
8. 435 rpm
9. 200 rpm
10. 123.75 rpm

Mean (Average), Median, and Mode, pp. 68-69:

1. 3,019 rigs
2. 43.5¢ or 44¢/copy
3. 40.91
4. $1,860
5. 12″, 1′, or 30.48 cm

Tables and Graphs, pp. 74-77:

1.

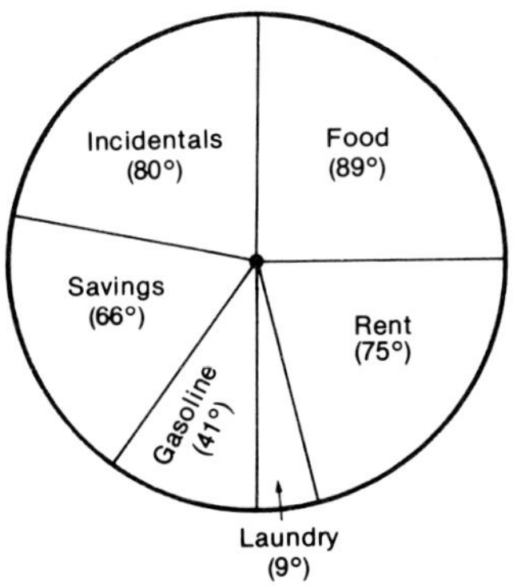

2. a. September, 1984
   b. Increased
   c. 8 million bbl/day
   d. 44 million bbl/day
   e. 1981

3.

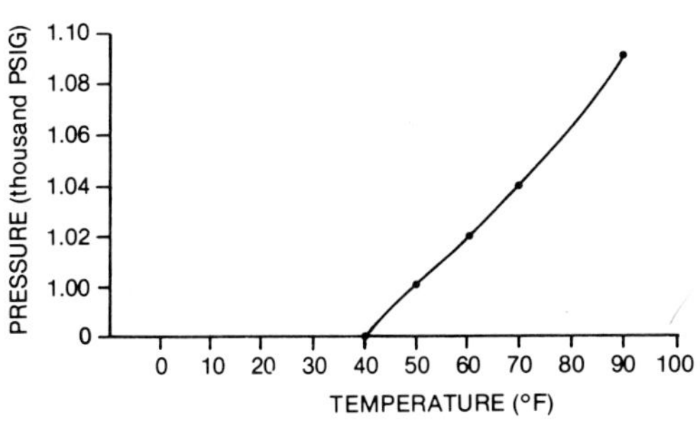

4.

GAS PIPE INSTALLED BY TYPE

Year	Plastic Pipe (mi)	Steel Pipe (mi)
1975	15,985	12,663
1976	11,640	9,718
1977	15,991	6,157
1978	14,576	7,212
1979	18,826	8,145
1980	18,912	8,334

5.

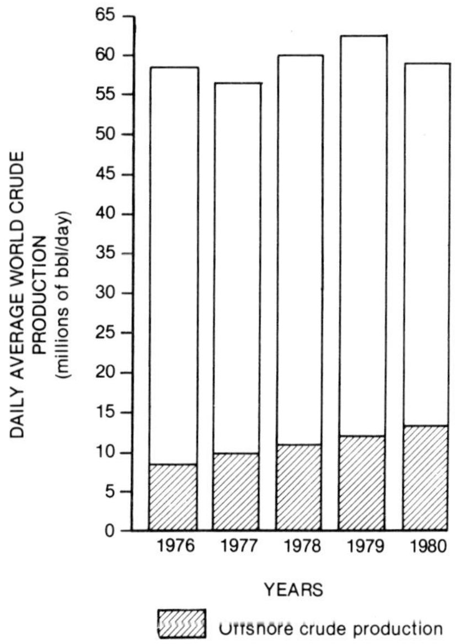

4. SOME PHYSICAL QUANTITIES AND THEIR MEASUREMENT

Length, Area, and Volume, pp. 94–95

1. a. $3/8$ in.
   b. $5\,3/4$ in.
   c. $1\,1/2$ in.
   d. $3\,1/16$ in.
   e. $3\,1/4$ in.
2. a. 7.4 cm
   b. 3.8 cm
   c. 3 cm
   d. 1.2 cm
   e. 7.9 cm
3. 91.44 m
4. 30.48 cm
5. 51.49 km
6. 1.93 m
7. a. 2.40 m
   b 2,400 mm
   c. 0.0024 km
   d. 2,400,000 μm
   e. 0.00000240 Mm
8. 1,196.17 yd^2
9. 14 hr
10. 10,857 $in.^3$
11. 144 $in.^3$
12. 4 bd ft
13. 1.48 yd^3
14. 0.00156 mi^2
15. 320
16. 1.307 yd^3
17. 158.928 litres
18. 7.085 ft^3
19. 1,308.64 acres
20. 2.396 mi^2

Time and Temperature, pp. 99–100:

1. a. 530 mph
   b. 8.833 mi/min
2. 21 min
3. 1,350 rpm
4. a. 161.6°F
   b. 68°F
   c. 32°F
   d. 22°F
   e. −108.4°F
5. a. −67.2°C
   b. 0°C
   c. 4°C
   d. −33.3°C
   e. −263°C
6. a. 51°C
   b. −70°C
   c. 20°C
   d. −23°C
   e. −224°C
7. a. 1,140°R
   b. 463°R
   c. 503°R
   d. 436°R
   e. 347°R
8. 12:47 P.M., Wed.
9. 99 min
10. 180.5 K

Mass and Related Derived Quantities, p. 108:

1. 907.185 kg
2. 53,376 daN; 0.533.76 kN
3. 631.579 ft
4. 11.023 lb
5. 1.8 lb/ft^3
6. 40.7669°API
7. 3,200 ft-lb
8. 51.6 psia
9. 15.92 lb
10. 6.928 psi

Electricity, pp. 113–114:

1. 0.12 A
2. a. 60 V
   b. 3 A
   c. 12 V
   d. 3 A
3. 172.5 kWh
4. a. 1.96 Ω
   b. 0.385 A
   c. 0.256 A
   d. 19.2 V
   e. 148 Ω
5. 180 W
6. 240 Wh
7. 4 Ω
8. a. 7.5 A
   b. 225 W
9. 0.5 A
10. 500 W

5. PRINCIPLES OF ALGEBRA

Algebraic Expressions, p. 125:

1. $3a - 3b$
2. $11a + 3b - bc + 6c$
3. $2a - 4ab + 26b - 2c$
4. $-3x + xy + y$
5. $15a - 8ab + 8b + 9c$
6. $a^2 - 4b + c$
7. $4x^2 - 6x - 6xy - 7y$
8. $-4a + 3ab - 15ac - 12b - 8c$
9. $5a^2 - 3ab - 14c^2$
10. $6x^2 - 10x - 4xy - 6y$
11. $4ax^2 + 8axy + 4ay^2$
12. $20a^2b + 30a^2b^2 + 15abc$
13. $3a^2 + ab - 4b^2$
14. $-4a^3 + 4a^2 - 8a^2b + 12ab - 4ab^2$
15. $x^2 + 2xy - 2x + 3y + y^2$
16. $-4a - 2b$
17. $9a^2 + 3a + 4b + 2b/a$
18. $m^2 - 3m + 14$
19. $x^2 - 2xy + y^2$
20. $m^2 - 4m + 8$

Equations, pp. 127–128:

1. a. 36
   b. 22
   c. 10
2. 2
3. a. $a + c + b = e + a + b + d$
   b. $b + c + e = b + d + 2e$
   c. $c + a + d = c + e + b + d$
4. $71.50 per day
5. 3.36 lb
6. Greater number: 20
   Lesser number: 15
7. 1,225 ft^2
8. $147.69
9. 90
10. $5.00

Formulas, pp. 135–137:

1. 0.4 darcys
2. a. 0.2364 in.
   b. 94.392 mi
   c. 6.564 yd
   d. 11.82 in.
   e. 6.562 ft
3. $130.00
4. 3.68 gal
5. a. $D = ST$
   b. $S = D/T$
   c. $T = D/S$
6. 55,063, or approx. 55,000 bbl
7. $SG = 141.5 \div (API + 131.5)$
8. 38.76 °API
9. $P_1 = P_2V_2 \div V_1$
   $V_1 = P_2V_2 \div P_1$
   $P_2 = P_1V_1 \div V_2$
   $V_2 = P_1V_1 \div P_2$
10. 1,651.26 ft^3

6. PRACTICAL GEOMETRY

Plane Figures, pp. 153–155:

1. 330 ft^2
2. 4.5 ft^2
3. 42 in.
4. 37.19 acres
5. 1,111.11 yd^2; 136.66 yd
6. 25 lb
7. ¾ of an acre
8. a. 11,200 ft^2
   b. 1,400 boards
9. 73.9, or 74 ft^2
10. a. 38.0431, or 38 ft^2
    b. 9 $in.^2$
    c. 65.96, or 66 $in.^2$
    $b = 8.25''$
11. 37.70 $in.^2$
12. 179.93, or 180
13. 0.0865 of an acre
14. 15 $in.^2$
15. a. 212.06 $in.^2$
    b. 805.82 $in.^2$

Solid Figures, pp. 164–168:

1. 10,378.125 lb
2. 138.88 yd^3
3. 7.48 gal
4. 1,468.698, or 1,469 gal
5. 204.75 bbl
6. 7.83 gal
7. 24 lb
8. 10 ft^3
9. $3^1/9$ yd^3
10. 12.1 ft^3
11. 11.13 lb
12. a. 18.96 ft^3
    b. 26.4 times/min
13. 4.14 or 5 fans
14. 20.23 lb.
15. 71 ft^2

Geometric Constructions, pp.178–180:

Since these are construction exercises, no answers are given.

7. TRIGONOMETRY

Right Triangle Trigonometry, pp. 195–196:

1. 75°
2. 8.48 in.
3. 42.4 ft
4. 8 in.
5. 6.93 in.
6. 20.4 in.
7. 15.05 ft^2
8. a. 22.3 ft
   b. 20.1 ft
9. 41°24″
10. 4.3°

Oblique Triangle Calculations, pp. 199–201:

1. 81.91 ft
2. 9.25 in.
3. 58.40°
4. 5.63 in.
5. X = 0.88 in.
   Y = 1.52 in.
6. 2.39 in.
7. 15.05 ft^2
8. Approximately 300 ft^2
9. 13.5 ft
10. 3.53 in./side

8. ADVANCED OIL INDUSTRY APPLICATIONS

Drilling, p. 221

1. 1,206 sacks
2. 173 bbl
3. 299 ton-miles
4. 285 ton-miles
5. 4,212 psi
6. 11.7 ppg
7. 1,325 psi
8. 10.9 ppg
9. 0.52
10. 625 psi

Production, p. 229

1. 0.75 bpd/psi
2. 14.06 hp
3. 4.03 pt
4. 38.59 Mcf/hr
5. 1.54 MMcf

Pipeline, p. 236

1. 1,732 ft
2. 478.32 hp
3. 81.1 lb/ft
4. ½″
5. 8,805 ft

Refining, p. 240

1. 0.813 Btu/lb/°F
2. 81%

Answers to Self-Tests

1. THE NUMBER SYSTEM

pp. 34–38:

1. 2,242,000,000
2. 19/40
3. 27/32″
4. 4 1/8″
5. 27/32″
6. Royalty owner: $19,880
   Operator: $139,160
7. 1,837 gal/hr
   44,100 gal/day
8. a. 0.5625
   b. 0.0938
   c. 0.4844
   d. 2 21/32
   e. 1/16
9. 610.3 mi
10. 152.4 bbl
11. a. 266.4 lb
    b. 177.6 lb
12. $2.11
13. 64 pins
14. 4.7465 in.
15. a. $190,000
    b. 6,552 bbl
    c. 26,208 bbl
    d. $464
    e. 409 or 410 days
16. 533 1/3 bbl
17. a. 11
    b. 348
    c. 71,639,296
    d. 6.346
    e. 10,000
18. a. 0.109 in.
    b. 0.133 in.
    c. 0.154 in.
    d. 0.203 in.
    e. 0.216 in.
19. a. 11,250 gal
    b. 236,250 gal
    c. 337,500 gal
    d. $6,851.25
    e. $2,500,706.25
20. a. 3,063
    b. $8.50
    c. 2.75
    d. $9.27
    e. 6,400

2. THE CALCULATOR

pp. 55–56:

1. 45
2. 33.4411
3. 31.88
4. 8.33
5. 1
6. 365.6
7. 14,300
8. $2.00
9. 2,329.6
10. 1.12
11. 14.213986
12. 436.8
13. 2,496.7091
14. 3.898885
15. 59.16567
16. 12,016
17. 74.934007
18. 14.3175
19. $71.70
20. 2,303.818

3. NUMBER RELATIONS

pp. 78–82:

1. 1 to 4
2. 312.5 ft
3. 1:3
4. 14.28 gal
5. $576.00
6. 3.66 hr
7. 480 rpm
8. 294.54 rpm
9. 200 rpm
10. $40.71/bbl
11. 128 ft
12. Mean: 51.62°F
    Median: 52°F
    Mode: 52°F
13. $2.65/Mcf
14. 7.39¢/kWh
15. 39 qt

16\.

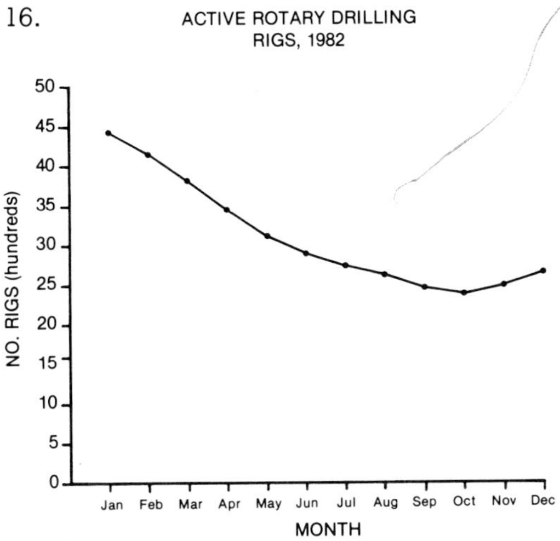

17\.

U.S. WHOLESALE PRODUCT PRICES ($/bbl)

	OCT. 1982	NOV. 1982	DEC. 1982
Gasoline	$39.68	$38.87	$36.88
Kerosine	42.20	42.97	36.88
Fuel oil	40.06	40.79	37.84
Residual fuel	24.78	26.06	24.57

18\.

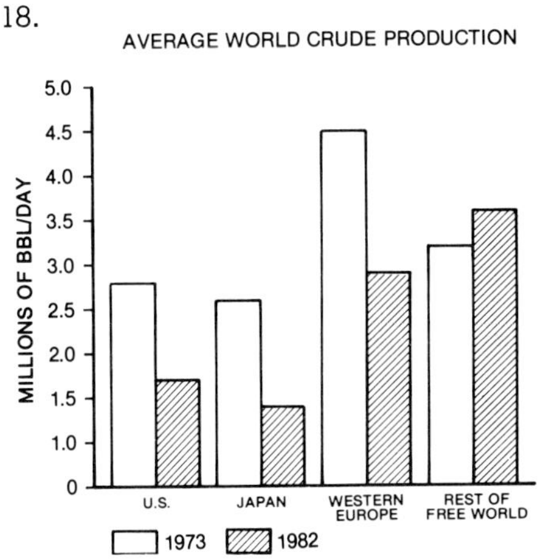

3. NUMBER RELATIONS, CON'T.

19.

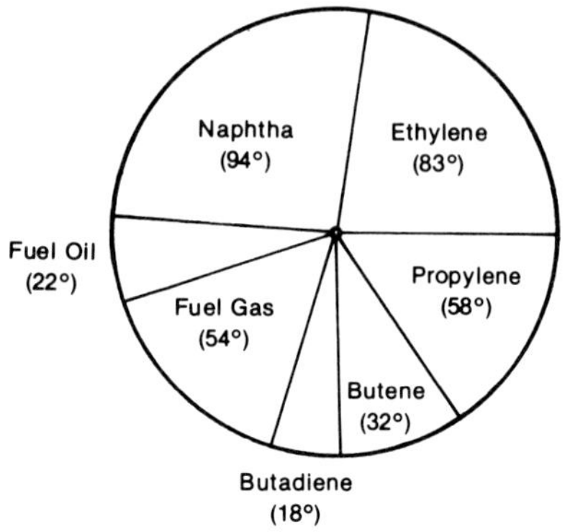

20.

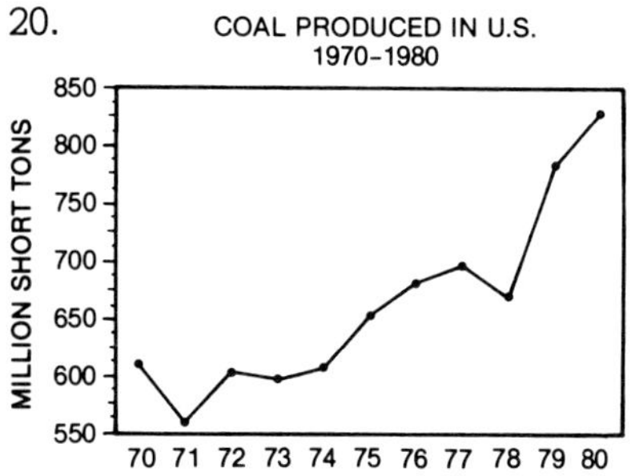

or

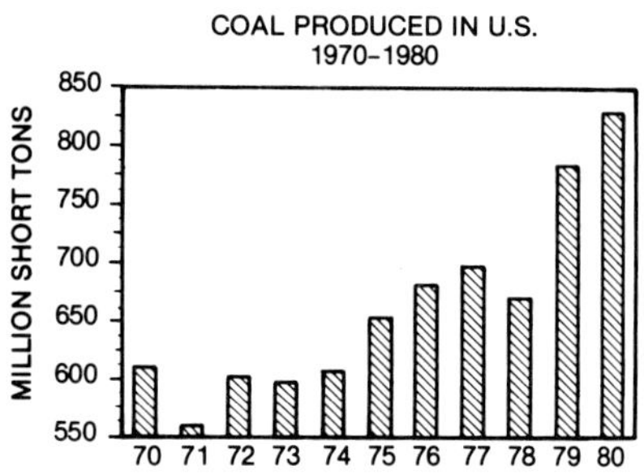

4. SOME PHYSICAL QUANTITIES AND THEIR MEASUREMENT

pp. 115-118:

1. 6.21 mi
2. 79.45 m^3
3. a fundamental quantity
4. a. mi
   b. lb/gal, or ppg
   c. ft^3
   d. m^2
   e. lb/in.^{2}, or psi
   f. g/cm^3, or g/cc
   g. lb_f
   h. Ω
   i. ft/sec
   j. rpm
   k. kWh
5. a. 10^3, or 1,000
   b. 10^{-1}, or 0.10
   c. 10^{-3}, or 0.001
   d. 10^{-6}, or 0.000001
   e. 10^6, or 1,000,000
6. a. 29.03 kg
   b. 3.5816 tn
   c. 17.984 lb_f
   d. 165.345 lb_m
   e. 2.222 oz
7. 25 pF (picofarad)
8. 26.25 lb/m
9. 268.56 MJ
10. a. 21.1°C
    b. 51°C
    c. −18.4°F
    d. 365.4°R
    e. 612°R
11. 7.52
12. 335,000 lb
13. 12.08 ft^3
14. 21,600 ft-lb
15. 279.75 kW
16. 10 A
17. 2.5 V
18. 6.93 psi
19. a. 570°R
    b. 296 K
    c. 53.7 psia
    d. 30.7 psia
    e. 1,760°R
20. 347°R or −112.2°F

5. PRINCIPLES OF ALGEBRA

pp. 138–142:

1. a. $x = 4$
   b. $x = 9$
   c. $x = 30$
   d. $x = 20$
   e. $x = 16$
2. a. 45
   b. 66
   c. −78
   d. 54
   e. 107
3. 1,500 lb
4. a. $9x - 3y$
   b. $-x - 4y$
   c. $20x - 5y - 7z$
   d. $a + 5b + 8c + 21d$
   e. $x^2 + 11x^2y + 8xy^2$
5. a. $-a - 4c$
   b. $-3b - c$
   c. $x - y - 4z - 6$
   d. $3a^2 - 7a^2b + 2ab^2 + b^3$
   e. $8x^2z + 9z + 2x^3 - xz^2$
6. a. $7a + 2b - 4$
   b. $2x^2 - 3x + 12$
   c. $3x - 12y$
   d. $9a - 7b + 6c$
   e. $-2y^2 + 4a^2$
7. a. $-5ab^2$
   b. $6x^5$
   c. $-24x^8$
   d. $4x^3y + 8x^2y^2 + 4xy^2$
   e. $-x^5ay + 2x^4a^2y - x^3ay^2$
8. a. $8a$
   b. $-3x^3$
   c. $x/5$
   d. $1 - 2(a + x)^2 = 1 - 2a^2 - 4ax - 2x^2$
   e. $a^2 + 2ab - b^2$
9. 140 acres
10. 20 ohms
11. a. $A = 118.30$
    b. $h = 16.77$
    c. $d = 10.69$
    d. $A = 31.82$
    e. $P = 23.55$
12. Larger part: 36
    Smaller part: 19
13. 84 ft/sec
14. 57.25 mph
15. a. Plan A: 33¢; plan B: 25¢
    b. Plan A: $2,994; plan B: 4,286
    c. Plan A: 12,006; plan B: 25,714
    d. Plan A: $25,000; plan B: $52,500
    e. Plan B costs $2,000 less.
16. °R = $^9/_5$°C + 492
17. $w = A/l$
18. 8 ft
19. 289.15 lb
20. 108 tiles

6. PRACTICAL GEOMETRY

pp. 181–185:

1. 12.5664 mm
2. 0.3314, or 0.33 $in.^2$
3. 1,368.84 gal
4. 150.8 $in.^2$
5. 1,436.76 $in.^3$
6. 1024.43 bbl
7. 18 $in.^2$
8. 16.98 acres on right; 144.63 acres on left
9. 1.2, or $1\frac{1}{5}$ gal
10. 27,073.54 gal
11. 563.98 gal
12. a. 97.23 ft
    b. 7.57 ft
    c. 1.55 ft
    d. 11.66 mm
    e. 1.658 in., or $1\frac{7}{10}$ in.
13. 131.9472 + 1, or 133 bolts
14. 534.38 ft^2
15. 1.2, or $1\frac{1}{5}$ gal
16. 105 $in.^2$
17. 2.25 ft^3
18. 448.18 rpm
19. 37.70 $in.^2$
20. 4.97 $in.^3$

7. TRIGONOMETRY

pp. 202–206:

1. 17.3 ft
2. a. 39.3 ft
   b. 22.6 ft
3. 46.7 ft
4. 5.6 in.
5. 12.1 ft
6. 591 ft
7. a. 619.3 ft
   b. 385.5 ft
   c. 143 ft
8. Angle A: 10°
   Angle B: 80°
   Angle C: 90°
9. 7.3 ft
10. 8.7 ft
11. $17'1\frac{7}{8}''$
12. Side a: 156.2 ft
    Side c: 294.8 ft
13. 22′3.5″
14. a. 1.2349
    b. 0.8098
15. 3 in./side
16. 2.3 in./side
17. 13.1 in.
18. 2.5 in.
19. 2.3 in.
20. 9.9 in.

8. ADVANCED OIL INDUSTRY APPLICATIONS

pp. 241–242

1. 11.2 ppg
2. 12.27 psig
3. 1,615.54 psi
4. 3.08 MMcf/d
5. 163,800 lb
6. 2 P.M. on Wednesday
7. 65 psi
8. 1,873 ft^3
9. 1,873,500 Btu/hr
10. 2,853.76 ft^3